Copyright © 2017 by Al Barrera

First Edition
First Printing, 2017
ISBN 978-0-9909432-4-2

www.al-barrera.com

The

Ghosts We Leave Behind

A Note from the Author

Many things have been said about the war in Iraq, and much more needs to be said. From our involvement to our justifications, you can't throw a rock without hitting a book about it. Some are historical accounts. Some are political step-by-steps guiding readers down the road of how we got there as best an author is able. There have been many amazing books about singular events in the war, from big battles to the deaths of significant figures.

This book is none of those things.

What I set out to create was a record of sorts, but not a perfectly historical one. While everything in this book is true, I have changed some details. Names are the most obvious of those. Some names changed, some didn't, but getting caught up on who's who is to miss the point. This story could very well be that of any of the thousands of young men and women who served overseas during the war, and I wanted to preserve that idea. Beyond that, I fudged the order of things a bit. I switched up who did what in an effort to make the story more coherent and reduce the number of characters down to something manageable.

Don't sweat all that. This book is about feelings. I wanted to explore what it was like to be a soldier during the Iraq War, and how it felt to see it all fall apart. The people and politicians of the United States seem to do their best to pretend Iraq never happened. ISIS and other extremists make it difficult, but I think we as a society are still giving it the old college try. Recent events with our elected officials are bringing a lot of this back to the fore, and it's as good a time as any to make my voice heard on the subject.

To see all the hard work we put in go to waste. To wonder if it was ever right to begin with now that I have the fortune of hindsight. To watch the politicians who so bravely told their people it was a war of necessity now try to disentangle themselves from the destruction they mandated. Being a veteran after a war is mentally and emotionally complicated, and this book is my attempt at unpacking that.

So, no, it isn't necessarily a history book; it's a book of the mind. I want to try to help people who weren't there understand how some veterans feel looking back on it. I'm hoping to spread awareness of the struggles we face because of our actions, and in doing so, I'm hoping to find a little closure on this chapter of my life.

Keep your seatbelts on; it's a weird ride.

A Quick Note About the Military

Before we jump in, I wanted to give a brief overview of several concepts that civilians might have a hard time wrapping their heads around.

The first and most obvious is army rank. The two most popular paths of progression in the army are officer and enlisted. Both start out at or near the lowest end of the scale and work their way up with time. (And in theory, merit) On the enlisted side it goes from private, to specialist, to sergeants. Sergeants are your buck sergeant, to staff sergeant, and finally, sergeant first class. After that you move into senior leadership with master sergeant, first sergeant, and sergeant major. Officers go from lieutenants, to captains, to majors, to colonels, and ultimately into the various types of generals.

The lower end of both spectrums are the rank and file of the military. Officers generally fill an administrative and second tier oversight role. Sergeants, also called Non-Commissioned Officers, are the backbone of the military and take care of most of the direct oversight on nearly every level.

How the army divides units will depend on where you are. In Charlie Company, two four or so man teams made a squad. Three to four squads make up a platoon, and three to four platoons make a company. Several companies make a battalion, and several of those make a brigade. Beyond that you move into larger unit sizes which won't be relevant to this book.

In fact, most of what I just told you won't, but I want to make sure civilian readers have at least a marginal understanding of how the army is structured before we jump into the good stuff. Battalion might get mentioned once or twice, but company and squad will be most of what you hear about here. As far as rank, just remember that sergeants are above privates and specialists, and officers are above them.

How we acted and spoke is as intact as I could make it, so the book will contain a lot of military jargon. There are footnotes to help you get through if you aren't familiar with the terms, so don't be too worried about getting lost.

And with no further input from me, lock and load. Keep your eyes up and your weapon ready.

If I die in a combat zone, box me up and send me home,

Pin my medals on my chest, tell my mom I done my best,

Tell her when she goes to sleep, remember me but don't you weep.

— Military Cadence

Chapter 1

"What the fuck are they going to do if I'm late? Send me to Iraq?"

He'd said it to be funny, but his boss had no sense of humor.

"All right," Staff Sergeant Miller said. "You can just eat a fucking MRE[1] in the truck."

Benjamin shoved the government cheese into his mouth, forgoing the hard-as-a-rock bread. He could taste all five years it must have sat on that shelf in the connex, forgotten by the world until they'd pulled the box out and thrown it on their truck.

Nobody spoke. In the darkness of the Stryker at night, only the glow of the FBCB2[2] and blinking green lights gave any indication of illumination in the world. The roar of the engine and the whine of gears filled the space, echoing back on them. But outside, it would have been quiet. Ghosts, the Iraqi people called them. Silent in,

---

[1] Meal Ready to Eat
[2] Force Battle Command Brigade and Below: Unit wide GPS and troop management system

silent out. You never knew they were there until they kicked in your door.

But there would be no door-kicking tonight. Just a patrol. Tooling around the city for hours in the dark, waiting for something to happen. They called it presence. Keep it random and hard. Make the enemy scared of what might happen if they decided to do something. Make them afraid to step out of line. Deny them any opportunities.

He called it boring.

Simmons broke the silence first. "You sure you need all that cheese, fatty?"

Masters, who had been resting his head against the gun rack, sighed. "Here we fuckin' go."

Benjamin savored the cheese and stared at the silhouette of the little man before him. When he spoke, he did so in the same lisping Boston accent as Simmons. "I dunno, Sthimmons. You think your sister will still bwoh me if I put on a few pounds?"

Masters laughed from the seat next to him. Donner joined in over the headset.

Simmons just stared. "Keep making fun of that one thing, 'cause that's all you got."

Benjamin shrugged. "Quit sticking your dick in that bear trap."

They hit a bump, and everything jumped into the air. You could say a lot about Strykers, but you certainly felt every blemish on the road.

The radio burst to life with a sudden crackle loud enough to drown out the constant chatter of the machines around them. Staff Sergeant Nelson in the vehicle behind them came over the speakers behind Benjamin's head and inside his helmet. "Sutter looks like he's assed out in your hatch, 5-2. Slap him."

The *beep* of the radio being keyed played over the headset. "I'm not," Sutter said.

"You better fucking not be, Sutter," Miller said.

"I'm not, Sergeant."

Benjamin piped in, "Yeah, Sutter. Wake the fuck up."

"I'm not, *Ben-ja-min.*"

"Benjamin." Miller used his no-bullshit voice, the only one he had.

A split second later, a foot caught the packet of cheese between Benjamin's hand and his helmet, stomping it everywhere. Simmons burst out laughing.

"Goddammit." Benjamin wiped the cheese away.

"Did I get him?"

"Yeah, you got him, Sergeant," Simmons said.

"Good, now shut up. And stay the fuck awake."

They all stopped speaking. Aside from the occasional squawk of the radio and ensuing chatter, the vehicle's noises consumed all the other sounds in the world. The grinding as they turned. The thud as they hit bumps in the road. It was a cage, one that kept them safe.

They sat and waited. Everything here was waiting. Wait to go on a mission. Wait to find an enemy. Wait, wait, wait. Only a few months in, and purpose had become a hazy concept. Something on the edge of sense or reason. No logic. Just a war zone with infinite time between it and home.

Benjamin rested his helmet against the Stryker as he cleaned the mess. So easy to forget a hostile country lurked on the other side of all that metal. Bombs and bullets with their names written all over them. A legion of people waiting for their chance to spill American blood.

Easy to forget now, but it had been a nightmare when they'd first arrived in country.

The plane had thudded onto the runway. They'd taxied across the tarmac to the far side of the airfield before the doors lowered and a blast of heat rushed in. One hundred thirty or so of the toughest sons of bitches he'd ever known stared out. Ready. Expecting. Half of them had been there before. Some had been training for this moment for years. Some, like Simmons and Masters, had only arrived at the unit months before.

It didn't look like much, but his stomach danced as he gazed out over the dusty base. The netting between connexes and wooden buildings. The huge T-barriers[3] that compartmentalized everything so that when a mortar or rocket landed it wouldn't do too much

---

[3] Texas Barriers: Tall, upside down T shaped barriers made of cheap concrete.

damage. The expressions of the people coming out to greet them. No smiles. No humor. This was war, and war was serious business.

He would never forget the look on Nelson and Sergeant Duarte's faces either, even if he couldn't quite place it. Was it defeat? Acceptance? What Duarte said to Nelson as they stepped off the plane only drilled it home further.

"Goddamn. Gone for two years and it feels like I never left."

Nelson only nodded.

The rest of the day flew by in a flash. Duarte patting Benjamin on the shoulder as they stood in formation, yelling over his head, "Welcome to Iraq, new guys."

The explosion in the city as they stood there at attention. The veterans in the company cheering as his stomach dropped.

"Was that an IED?" he asked Nelson.

"Fuck if I know."

Actions change the future, not hoping. Not sitting with your thumb up your ass. Getting out there and doing the hard deeds of war. Soldiers understood that better than most. But hearing that explosion out there, seeing the fear in the eyes of the men around him, the excitement in others, he wondered if he hadn't made a bad call signing up.

Captain Clinger and First Sergeant Ferrelo spoke to them. Psyching them up. "This is it, men. You're fucking killers. Don't forget that shit."

They spent the rest of the night and the following week unpacking, getting ready, and transferring vehicles. The riflemen did left and right seat rides with the unit they were replacing. New SOP[4] came down every day.

Soon, the CP[5] was set up, and the mortars were on guard or lifting boxes instead of doing the job they'd trained to do for years. Watching the MWR[6] tent. Pulling radio watch when they weren't running paperwork and lifting boxes for supply.

He'd been happy about it at first. Why the hell would anyone want to go out and get shot at? Before that deployment had come down, he'd been a hardened killer. But once they were going, once they were on their way, once they were there, he'd wanted nothing to do with it.

At first.

The half-remembered nightmares that had started when they'd arrived in country tapered off. His friends came back with stories about the thrills and dangers of patrols. Raids. How it felt to be a soldier, to be the men they had trained to be for years now. Together.

The base became a prison, though it was nice enough. The CP was just a wooden shack. The chow hall served better food than they had at home. Same with the big, tan MWR tent. Nicer than their rec room back at Lewis and a hundred times the size. The CHUs[7] they

---

[4] Standard Operating Procedure
[5] Command Post
[6] Moral, Welfare, and Recreation
[7] Containerized Housing Unit

lived in were little more than a connex room made for two. The PX[8] sold everything they needed, and the phone center wasn't far.

Still, the countless T-barriers and sand bags never let him forget where they were. It didn't look like he'd imagined a military base overseas would. Everyone had a weapon, of course, but it was more haphazard than he'd expected. Secure as a bank vault but not as nice to look at.

The explosions in the distance, the gunshots. The endless parade of soldiers everywhere and vehicles always driving by. Civilian contractors walking around as if they owned the place. The dust. The call to prayer in the distance. He wanted out. Specifically, he wanted to be out there with his brothers.

He started to feel less and less like a member of the group. Just an appendage now. Not a killer, just the guy that held the purse. Saluting generals when they visited. Standing in a tower and staring out at nothing but a dusty road and buildings in the distance. Arguing with the other mortars because they were just as edgy for the same reasons.

"You should be glad," Nelson said to him. "It's not cool out there."

Easy for him to say. Nelson was a mortar too, and one hell of a squad leader. That was exactly why he'd been picked to fill in on another platoon while headquarters platoon got the shit end of the stick, as usual.

---

[8] Post Exchange: Military store

Until one night, they didn't.

Miller barged into their CHU. "Get your gear ready. You're coming on patrol tomorrow."

Benjamin pulled his headphones off and sat up in bed, his heart skipping a beat. "Really?"

Miller wore Coke-bottle glasses that transformed every expression into something cartoonish, his, "*What the fuck did I just say?*" expression most of all.

Benjamin took the cue. "Roger, Sergeant."

He was out the door a moment later, no details given.

"Really?" Sutter said in a mocking voice, and they both laughed. The moment they'd waited for had finally come.

They were soldiers.

Within two weeks, command formed fourth platoon. They threw mortars, the tankers, and one rifleman squad together to put more boots on ground. The details and the bullshit duties kept coming, but that was to be expected. They were in the army, after all, but no longer held prisoner on the FOB[9]. They'd found the place they'd always wanted to be. The place they'd trained to live.

The *tat-tat-tat-tat* of an AK bursting somewhere nearby pulled him from his thoughts as he removed the last of the cheese from his helmet.

"Shots fired, right side," Nelson said over the radio.

---

[9] Forward Operating Base

It was probably nothing. After only a few weeks of going out the wire, gunshots became mundane. Just another hadji poking a rifle around a corner and shooting without looking as the Strykers passed by. Maybe someone taking pot shots from a rooftop a quarter mile away.

He wondered if their enemies saw more purpose in all of this than he did. The country belonged to them, after all.

"See anything, Sutter?" Simmons asked.

"I don't see shit."

They almost never did.

The patrol ended, and they returned to the FOB. There was only enough time for food and sleep before it was back out again for QRF[10]. Sometimes it felt like there was no free time, and others it felt like it was endless. Never had he lived more in the moment. How could you not? To try anything else at war was reckless, if not for your body then for your mind. There were certainly endless conversations about home. What they would eat. Who they would fuck. The things they would do. But it was always distant. They had a long way to go.

The trucks pulled into the old Iraqi army base they used for QRF. Resolve. In truth, it was an unfinished building, five stories tall, surrounded by T-barriers. Beat up and ragged, like most things in Iraq. Long disused trailers sat in what had once been a parking lot. Military vehicles, piles of trash, potholes, and dust. Always dust.

---

[10] Quick Reaction Force

Buildings all around them formed the heart of the city just outside the walls. The soldiers there barely spoke a dozen words of English between them, but that was what the terps[11] were for. A MiTT[12] team, a handful of American guys who lived in a more secure section of the building, remained there year around to train the Iraqis. One was from Charlie Company like Benjamin and the rest. It was always good to visit when they had to wait there in twenty-four hour shifts for something to happen. That was all QRF was. Waiting.

No sooner had they pulled in than a bomb went off in the city, a distant boom not close enough to be felt, but not too far off either. A month ago, it would have made Benjamin's heart beat faster. There would have been adrenaline and apprehension. Now, he only sighed.

"Fuck," Simmons said under his breath.

Sergeant First Class Garcia, the tanker turned infantry platoon sergeant, spoke up over the radio. "Don't bother unloading. You know we're going to get called to that shit."

They dropped the ramp so daylight could spill in while they waited for the inevitable call. There was no air-conditioning in the Strykers. Having the ramp down at least allowed in a breeze that made the blistering summer heat more bearable. Benjamin, Simmons, and Sutter stepped out to smoke while Miller dropped from his hatch and plopped down in his seat. Donner didn't bother getting out of the driver's hatch.

---

[11] Interpreter
[12] Military transition Team

"That shit's going to kill you. You're too young to be ruining your bodies like that," Miller said. It was much easier to hear in the Stryker when the ramp was down. It gave the noise somewhere to go.

"Roger, Sergeant." Benjamin took a drag.

"Don't 'Roger, Sergeant' me."

"Rog-o."

Miller opened his mouth to say something but only shook his head and muttered it under his breath. The others snickered.

Why the hell not smoke? Everyone wanted to ride their asses about something. People shot at them every day. It was one hundred twenty degrees out. When they took off their kit at the end of the day, salt lines had formed from all the sweat. The whole world smelled like shit, and not in some hyperbolic way. It literally smelled like burning garbage and human waste. Some spots stank worse, Resolve more than most. The war had transformed all of Mosul into a dump, and the whole fucking trash heap kept trying to kill them.

The battalion radio came to life. "QRF, this is Patriot Base. Over."

2nd Lieutenant Brenner answered, "Patriot Base, this is Charger 4-6, go ahead. Over."

"We populated the explosion and need you out there. Over."

"Roger, Patriot Base. Over."

"Patriot Base, out."

And just like that, they threw their cigarettes aside, put their helmets back on, and everyone piled right back in. The giant steel ramp rose into place, sealing their cage once more. Benjamin rubbed the sweat from his eyes. The days were hotter than balls, but they had light to see by, and they didn't have to wear goddamned NODs[13] on their heads. There was some small comfort in that.

Once they began to move, Simmons switched off the mic on his CVC[14] and side-eyed Miller to make sure he wasn't listening. He was too busy pontificating to Sutter about marriage while Sutter tried his best to sound anything but bored in reply.

"Did you know smoking was bad for you, Benjamin?"

"I like things that are bad for me. That's why I'm going to bang your sister when I get home."

Simmons smiled his half-smile and shrugged. "If you can get her, you can have her. But she's too much for you."

Benjamin shook his head. "I need DJ Buttercup in my life."

A helicopter, the one that always shadowed them in Mosul, flew by overhead, drowning out Simmons's laugh. "My brother is available too, if you'd rather go that route."

Benjamin placed a hand on Simmons's knee. "If I'm going balls-deep on a Simmons man, it's going to be you."

Simmons laughed even harder. Masters finally took notice of them. "What the fuck you faggots doin'?"

---

[13] Night Optic Device
[14] Combat Vehicle Crewman: Also used to refer to a vehicle helmet with a built in plug for internal communications.

Simmons cut off his laugh all at once, pretending to be angry at the interruption. "Nothing, blackie."

Benjamin rolled his eyes and turned his attention elsewhere. "Jesus, dude."

Masters licked his back tooth and shook his head, staring at Simmons with anger Benjamin could never place as real or fake. "All right. All right. I got you."

"I'm just kidding."

Benjamin, who was part Mexican and grew up with his black sister, had never found Simmons's humor funny, but Masters always blew it off. Such was the way of life in the infantry. Some managed to be hardened killers without giving up a piece of their tact and emotions, but they were few and far between.

"I know. I'm gonna be kiddin' when I stick my dick in your mouth while you sleepin' too."

Benjamin leaned his head back while they bickered, tuning them out. In moments, they reached the site of the explosion. The riflemen dismounted their vehicle a good distance away with the lieutenant, but there was nothing to see. A bomb had gone off on the side of the road. They found no ambush or secondary, just a small crater on a mostly empty street. People watched from behind fences, their long outfits and headdresses appearing as foreign as this place felt.

But the bomb? Nothing out of the ordinary. Yet somehow, they managed to draw it into a two-hour affair before they spun down and returned to Resolve. A microcosm of the world they now lived in. Explosion and violence all around, but no reason to get excited. Nothing to fight. Nothing to kill.

They unloaded and set up a rotating guard to listen to the radio, spending the rest of the night watching movies on the old television set in the big room on the third floor where the platoon slept. Benjamin pulled the middle of the night shift for radio guard with Ski—Kowalski—talking deep into their watch about the finer points of Pokémon and tank gunnery. The next day, third platoon replaced them and they returned to the FOB to get a few hours of sleep before the next detail or mission came down, the same thing day in and day out.

And even bored. Even hot. Even when things seemed bad, he wouldn't have it any other way. People might say they know, or pretend to understand. They would write the words as if it meant something, without the feeling, lacking the memory that placed them there when the bullets flew and the blood fell. But when they said, "These men, these are your family for the rest of your life. Closer than any sibling," they had no real idea what it meant. Just writers pretending, just actors poorly imitating the real thing.

This, though? He looked at Simmons as they climbed into the Strykers the next day. At Sutter and Donner. At third platoon as they filed into their vehicles, loading a round and checking their gear.

This was family.

This was war.

14

Chapter 2

Everyone had told him college was different, but they were wrong. It was just another kind of mission.

Miller had warned him that getting out was a mistake. "Nothing in the world is gonna compare to this," he said one day as the mortars ate chow together. "You'll regret getting out not a year after you do."

Benjamin belonged in the military with his friends. There were sheep, sheepdogs, and wolves, and Charlie Company came stacked with sheepdogs from top to bottom.

But that hadn't worked.

His mother had begged him to stay when he went home to Michigan to pack his things from his life before. But Detroit was the armpit of the United States as much as Iraq was the armpit of the world, and that hadn't worked either. So she gave him a gun for Christmas, helped him move, and sent him on his way.

He'd spoken to Allen Rosenberg on the phone as he pulled into his new life in Chattanooga in the car he'd bought with his Iraq money. His first car. Allen had been one of the third platoon guys he'd become friends with overseas.

"Welcome home, buddy! I'll see you in a bit."

He stayed with Allen for a few days, much to the chagrin of Allen's wife, Megan, who'd never met Benjamin. (Or "Jay", as he went by now. "Don't use my slave name.") But it hadn't been an issue for long. He found his own apartment and registered for college within a week. Nothing to do but wait out Christmas and start again in January. He and Allen passed the time drinking, celebrating their newfound freedom, and meeting Allen's friends, people who treated him as if they'd always known him. When they weren't drinking, they explored the valleys and parts of the city where Allen had grown up.

"There's so many fucking hills, dude." The highway cut through the split in the ridge going down I-75. Below them, Tennessee opened up in the valley like a promised land from on high.

"Yup, yup, buddy. We *do* live on a mountain or two."

Before long, the first day of school rolled around. His nerves jangled as he parked his car and grabbed the bag full of books that had been much more expensive than he'd expected. He could see his breath in the January air, but he was sweating under his jacket. He internally cursed himself. He'd survived a deployment to Iraq. He'd been around the world and done more by twenty-one than most ever would. There was nothing to be nervous about. It was stupid.

Dreams of being a firefighter had given away to ideas of being an EMT. EMT had turned into nurse, and nurse had become doctor. After all, if he was going to get educated, why not go all the way?

He walked around the big duck pond in the middle of campus that separated the parking lot from the cluster of buildings. His stomach in knots, he stopped to ask directions from a pretty blond woman smoking a cigarette outside of administration.

"Uh, excuse me?"

She looked up.

"I'm looking for the humanities building."

She pointed farther down the walkway without a word.

He thanked her and kept walking. He still had the bearing of a killer. Chest puffed out. Shoulders back. Chin high and head shaved. Was that why everyone kept looking at him? Did he really stand out that much from others his age who'd stayed home to play video games and smoke pot with their friends while he and his went off to war?

A crowd of people stood around the entrance smoking. They all appeared too young to be here. At one point in Iraq, he'd gazed in a mirror and failed to recognize the person staring back. He was still a kid by most people's standards, though he no longer felt like one.

He breezed past them and into the building. Art decorated the walls. A staircase led to a balcony above, and hallways spiderwebbed in every direction. He checked his food-stained and often folded syllabus for the tenth time. 201. He could have recited every piece of information on that sheet in his sleep.

His nerves rattled worse than before. Battle high. The feeling that shook you from the bottom of your feet to the top of your head when you knew a fight was coming. He took a deep breath and climbed the stairs two at a time. He wasn't afraid. It wasn't fear. He said it over and over in his head. There was nothing to be wary of here.

More tangled hallways filled the upstairs. He found the room and entered before anyone else arrived. Someone had once told him that students who were serious about their work always sat in the front, and he did just that. The minutes ticked by, and his classmates trickled in as his nervousness grew.

Not afraid.

Not afraid.

Some people in the room chatted, but most stared straight ahead like him, or gazed idly at phones half-hidden behind cupped hands.

Right on time, a little old lady walked in. The skin-colored pantsuit she wore hung off her as if she'd once weighed more. Her big glasses, not unlike Miller's, perched on the end of her nose.

"Good morning. I'm Mrs. Rebba, and I'll be teaching your humanities class this semester." She set a packet of papers down on the desk with a thud. "Don't worry; it'll be an easy day. I just want to go over the syllabus and tell you what I'll be expecting. How does that sound?"

Nobody responded.

She looked him right in the eyes and smiled. "Are you excited?"

To his surprise, he was. He smiled back. It wasn't fear. He wasn't afraid. This was a mission like the hundreds upon hundreds he'd run before. His heart soared. Tennessee was the right place, and school was the right choice.

"I am."

The class flew by as quickly as she'd promised, as did the next two, and the one on the next day. The whole week proceeded at a breakneck pace. What a change from the months in Iraq spent waiting to get on with life. The constant outer cordon and patrols. The indirect fire in the middle of the night. Life was supposed to be like this. Quick. Decisive. A sense of forward progress.

The following Saturday, he and Allen sat on the back porch of Allen's house drinking. Inside, Megan took care of the little girl born while her father was away. She was over a year old and living with her dad for the first time.

Allen took a shot of Old Crowe. "I'm going to breeze through these math classes. No sweat."

"Shit. I wish I could say the same. Math is the only thing I'm worried about."

They shivered as they smoked, staring at the cloudless January sky and the stars above. It was the only light in the world. So different from Michigan with its infinite street lamps and cars. Closer to Iraq than the place where he'd grown up. It would take some getting used to. All of it would.

"Math is just a language, man. If you think of it like that, you'll have an easier time than if you tried to memorize formulas."

"I suck at languages, too." Jay flicked his butt into the darkness and watched it sail away.

Allen followed suit. "You'll get it. You pretty excited, though?"

"I am." He slid the door open and stepped back inside. "But I feel like I'm in a pre-school with all the teenagers around."

Allen laughed as he plopped down on the couch and picked up his Xbox controller. "How the hell do you think I feel?" He was older than Jay by almost five years.

That was life for a little while, for both of them. Work, school, drinking, sleep, repeat. One weekend a month they went off to play army with some farmers who called themselves the National Guard, but it was nothing compared to the real thing. Allen went home to a family at the end of each night, and Jay went out to see the town.

It wasn't all roses, though.

Three months into the semester, the kids in the back of the world history class were grating on his nerves. The snickers. The sneers. The tight, oh-so-hip jeans with the bad attitude to match. They were babies pretending to be bears. The teacher, William McKee, was a no-bullshit old man and city council member. He was a vet too, though not of the same war.

20

But his smile didn't crack when those little bastards ripped into the current state of affairs with gusto. Jay's did.

"I'm just saying, George Bush is a terrorist. Anyone who fights for him is a terrorist, too." The nineteen-year-old kid with the My Chemical Romance haircut smiled as he spoke. "I know that isn't the cool thing to say, but whatever, it's the truth."

McKee eyed Jay, who sat in the front of class. That gaze said it all. *Keep your mouth shut.* Jay hadn't realized he'd been gripping the edge of his desk for dear life until he saw it.

"My Dad fought in Desert Storm. That war had a purpose. This thing going on now is just for oil."

It wasn't. His friends hadn't just died for oil and money. He was no far-right nut job, but you didn't have to be to see the left's agenda was wrong as hell.

"I'm just saying, there's—"

Jay turned in his seat, McKee's warning be damned. He could feel the veins in his neck popping out as he stared the kid down. "Why don't you just shut the fuck up?"

The whole room froze. Every eye turned on him, and he didn't give a damn.

"Hey, man, I—"

Jay pulled the dog tags out of his shirt. "I said shut the fuck up."

The kid's mouth opened and closed. Jay had seen the deer in the headlights expression so many times overseas. The fear and embarrassment that always came before someone slinked away with

21

their tail between their legs or did something very, very stupid. The kid's compatriots looked back and forth at one another, unsure of what to do when their leader had been called out.

William cleared his throat to bring the attention in the room back to him. "I think that's enough for today. Class dismissed." He gave Jay one last, lingering look that radiated disapproval before he returned to the podium and packed his things.

The nerve. The audacity. Free speech was all well and good, but that didn't mean there wouldn't be consequences. People couldn't just run off at the mouth as they saw fit. He'd earned the blood on his hands, bought and paid for it. Hell would freeze over before he let some punk who didn't know what his dick was for yet talk about the dead that way.

"Hey, man."

Jay had been busy shoving his books into his backpack, his pulse still pounding. He hadn't seen My Chemical Romance walk down the aisle or stop in front of him, but he saw the clenched fists now. The glare. His friends whispering to each other by the door.

Jay stood and picked up his backpack. If this kid was trying to look fierce, he'd have to try harder.

"You don't want to mess with me." The kid took another step forward.

The bag dropped from Jay's hand. His pulse quickened further. Not fear. Not in the least. Excitement. Anger. His fingers and toes tingled. "What the fuck are you going to do?"

And there it was, like every time before. The terms were on the table. This place wasn't so different from war. A little more civilized, a lot better smelling, but at its heart, it was just another type of conflict. He'd been forged in that every day for years in the military, training and patiently waiting until he'd put those skills to the test. He smiled, and when he did, fear darkened the young man's eyes.

William swooped in before either of them could move. He grabbed the young man by the shoulders and turned him around. "Let me speak to you for a moment."

The young man glanced over his shoulder at Jay as his friends stared from the door, doing their best to look defiant. But without a word, he acquiesced and left the room with William.

Jay's heart still raced as someone clapped him on the back. When he spun to see the smiling face behind him, he had to blink twice before it registered. Half the class remained in the room. He'd been so focused on the My Chemical Romance crew that he'd hardly noticed.

"That little tool's been getting on my nerves too. Good job." The kid who'd patted him smiled as he walked out. A young woman behind him smiled as she brushed the hair from over her eyes. Others looked at him and nodded as if he'd done something brave, though he felt like a fool the moment he left the building.

My Chemical Romance never spoke in class again. He didn't make eye contact, even when he saw Jay outside. When they both pulled into the same gas station up the road from Jay's apartment, the kid pretended not to notice. He looked at the sky, his car, the ground—anywhere but at Jay, who only stared holes into his skull.

He got into his car and pulled away without saying a word.

Chapter 3

The whirring of gears groaned over the radio chatter as the mortars turned down the street that would see them to the mission. A company-level operation. Supposedly, a bomb maker lived in one of these shacks at the edge of the city. So the ghosts would roll in quietly, ready to take him dead or alive.

Alive was always preferable. Despite all their talk of morals and a mission, most terrorists flipped on their friends the first time they took a fist to the gut. Nothing too serious. The line had to be drawn somewhere; that's what made one side better than the other. Grab a guy's nuts and squeeze them until he screamed, sure. But at the end of the day, they'd end up in a prison with all their closest jihadist friends, not on TV getting their head cut off. That separated the two sides in this war. Civilians could balk at the idea all they liked, but that's why they didn't know. They had no fucking clue how war worked.

Benjamin watched the landscape in the surreal green vision of his NODs as he stood out of the hatch, manning the 240[15]. The old world had flourished and faded here, giving birth to the new. Were they watching another one now? Another page in history that only seemed a footnote at the time?

It was hard not to be in awe of some of the things they'd seen. The walls of Nineveh, all that remained of that ancient kingdom, still stood. Sheep grazed on the side of the hill as they passed. On patrol, they visited the Tomb of Jonah. The final resting place of the man supposedly eaten by a whale when he refused to be a prophet. So many of the world's legends had taken place right here in this country.

And now it was a stinking, war-torn shithole.

"It's always been like this," Miller had said. "It's the natural way of these people."

But Benjamin didn't buy that. He'd grown up surrounded by Arabs and Muslims. A good chunk of the rest of the company bought into the bullshit always implied but never stated in the military, but he didn't. That these people had done something to deserve it. That they were all terrorists waiting to happen. They'd taken class after class about cultural awareness and sensitivity before they deployed, but little of it sunk in with most of the guys.

---

[15] M240B: Belt fed 7.62 machine gun.

"Look sharp, Benjamin." Miller turned in the hatch next to him. The NODs only made him appear more ridiculous than he normally did with his giant glasses. "Let Olson up."

He climbed down, glad to be done with it. Up or down, one wasn't more exciting than the other, just more likely to get you shot by some guy you never saw. The bomb maker wasn't going to come out guns blazing, if he was even there. And if he did, the riflemen would kill him before any of the mortars could say, "Shots fired."

At least in the truck he could sit. He flipped his NODs up as he sat in the back next to Sutter.

"Hey, buddy," Sutter said. "You kill anybody?"

Benjamin rested his rifle across his legs. "I'm about to."

Sutter laughed. "I seen you shoot. Safest place to be when you have a gun is right in front."

Donner laughed over the headset.

Miller cut it all off. "All of you shut the fuck up. You aren't fucking tourists, you're soldiers."

Benjamin rolled his eyes. No sleep. No purpose here. Just mission after mission without a hint of progress, and their boss still yelled at them as if they were storming the beaches of Normandy.

Sutter and Benjamin spoke quietly with the mics turned off, and the night passed slowly, as it almost always did. In the movies, soldiers' lives were filled with action. That hadn't been the case so far. When there was action, it was quick and violent. Most of the time spent at war thus far had been of the sit and wait variety. He

27

doubted it was different for the riflemen. Before long, they'd kicked in the door, found nothing but some scared women and children who didn't know dick, and were on their way.

It gave him a lot of time to think. A bomb maker. The kind of guy who lived to ruin your day in the most decisive way possible. Bullets didn't scare Benjamin. Snipers weren't magic. They had to see you to kill you. The IEDs[16] didn't. Someone set them up while you slept blissfully unaware that your death had been placed on the side of the road. Charlie Company had first deployed before he joined the army, but the guys told him IEDs hadn't been as bad back then. Now they had tank killers. They had EFPs[17] and infrared sensors. These people were smart, and anyone who disagreed was wrong as hell. Benjamin couldn't build an infrared tank-killing bomb, and he doubted anyone else in the company could either.

Would it hurt? Nelson had been hit with more bombs than anyone in the company. They'd scrambled his brains, and he forgot anything he didn't write down. He'd said it hurt like hell, that it rattled your teeth and knocked you out. Made you see spots. Shook your brain until it screamed in pain. It deafened you to any sound and choked you with the dust it kicked up, and that was only if it didn't pierce the hull or blow your truck to shit outright.

As if Benjamin's very thoughts had summoned it, a small explosion detonated a little ways behind in the middle of the

---

[16] Improvised Explosive Devices
[17] Explosive Formed Penetrators: A type of IED capable of penetrating any military vehicle.

company formation. The company and platoon radio came alive with chatter.

"Got us. Just a little one," Staff Sergeant Quinn said over platoon radio. Quinn's vehicle drove behind them in formation.

Lt. Brenner called it up over the company radio, and that was that. Just a little toe-popper, as his drill sergeant had once called them. Hardly worth mentioning. Underwhelming, like most things about war.

Two hours later, they'd made it back to base, changed the shredded tire, and forgotten all about it.

Within two weeks, he'd been made RTO[18]. That meant following the lieutenant around with a radio strapped to his back. Getting to dismount with the rifleman. Getting to be an honest to god soldier.

Miller had informed him in his classic manner, bringing it up in one of the countless squad meetings they had at the end of most days. "They asked me to pick someone to be RTO. Benjamin, you're it. Don't embarrass me."

The back of the infantry variant of Stryker was different from the mortars. A big 120mm cannon sat in the back of his old Stryker. The troop carriers, on the other hand, were made for transport. A whole squad fit in the back. Two teams of the hardest killers the planet had ever produced.

---

[18] Radio Telephone Operator

He rode with Quinn, Brenner, Specialist Jared, Specialist Howard, Sergeant Lunquist, and Sergeant Rosenberg now. The best men of the best platoon pawned off on fourth to fill out a roster, and they never let him forget it.

"Jesus Christ, Benjamin. You listen to Nickelback, don't you?" Quinn asked over the headset as they drove to Resolve for yet another QRF mission.

"No way, Sergeant."

Lunquist sat in the back, looking tired as he always did. "You and him can go jam out after this." He pointed at Brenner in the back hatch. "Maybe you two can croon love songs to each other after this mission."

"Yeah," Quinn said. "You guys can read your spoken word poetry before you blow each other."

They all laughed, including Benjamin. Killers were seldom nice men, though there were exceptions. None of it wounded. They spoke another language, no different from those who went to college. Their own way of communicating, much of it indecipherable to those outside.

They pulled into Resolve, and most of the men went upstairs to unwind and wait. Some strayed to the small kitchen the Iraqis kept downstairs. When they'd first eaten there, it had been lamb and rice soup with a side of diarrhea for days. As their bodies grew accustomed to it, the food became a welcome reprieve from the chow hall and MREs.

Benjamin pulled first radio watch, but it didn't last long before battalion called them up. Jared responded while Benjamin ran up the stairs to grab the others. Within minutes, they were loaded up, locked, and rolling. A flaming truck had been parked in the middle of an intersection and called in, and it was their job to investigate the obvious.

"This is a fucking trap," Quinn said over the headset as they rolled out of the gate they'd just entered. "There's probably a bomb in that fucker just waiting to go off."

"Yup," agreed Rosenberg.

Benjamin watched the sky through the open space in the back hatch between Brenner and the Stryker, the same sky he'd looked at a million times across his life transported to an upside-down world. "Are we just going to call EOD[19]?"

Brenner answered this time. "We'll see when we get there."

Lunquist and Howard caught each other's eyes for a moment. They didn't trust him; that much was no secret. He wasn't infantry. He was an armor guy who'd been stuck into an infantry role. While he'd spent his entire career alongside the queen of battle, that didn't make him infantry.

It didn't take long to find what they were looking for. Benjamin couldn't see it, but he didn't need to. Rosenberg pointed the .50 cal at it in the RWS[20], showing it on the screen for a brief moment before scanning past it.

---

[19] Explosive Ordinance Disposal
[20] Remote Weapons Station: A crew-level weapon mounted on a vehicle and

A truck burned in the middle of an empty road. That second part said everything they'd suspected. Insurgents seldom announced their plans, but sometimes word got out. A normally busy street empty in the middle of the day meant trouble.

"Yeah, let's just pull back and call EOD," Quinn said.

"Hold on," Brenner said. "We need to examine the site."

Despite all the machinery beeping and groaning around them as the truck came to a stop, you could cut the silence with a knife.

"What?" Quinn asked.

"We need to set up a TCP[21] and get a look at things from the ground."

"Sir," Quinn said in the most condescending way possible. "I think—"

But Brenner keyed the radio and made his orders known.

"This is fucking stupid," Lunquist said under his breath.

Rifleman or no, Benjamin was infantry. Step one of surviving was not stepping into the world's most obvious trap. He shifted in his seat and stared at Howard, who shrugged. Saying no was an option, but you couldn't prove a negative, and there would be consequences. A lot of words did less damage in the infantry than elsewhere in the world, but "no" wasn't one of them.

He'd been so busy thinking about it that he ignored the radio chatter. The next command was as familiar as breathing.

---

controlled electronically from within.
[21] Traffic Control Point

"Ramp clear." Brenner hit the small switch next to his hatch, dropping the two-thousand-pound steel slab.

The metal hit the pavement with a thud. Game faces on, weapons up. This was their work, and they did it well.

The street stood as empty as it had on the RWS. Nothing but the oddly square buildings the Middle East loved so much. A few beat-up old cars on the sides of the road. Dust and dirt everywhere, and the sun pounding as it always did. The only living souls in the world were the two children watching from the corner beyond the burning truck.

"This is fucking stupid," Lunquist repeated as he moved to step out.

He never got the chance. The platoon radio came to life. "Get the fuck back in your trucks. We ain't doin' this stupid shit," said Garcia.

An officer might be the one in charge, but nobody listened to an inexperienced officer over an experienced platoon sergeant. Lunquist stepped back inside, and they all took their seats.

No sooner had Lunquist clicked the button to put the ramp back in place than the first mortars fell.

Boom, boom, boom.

Pounds upon pounds of explosives and metal rained from the sky, hitting the intersection in and around the truck. Death from above. High-angle hell, as the mortars said, but these rounds fell for the wrong team. He caught one last look at the children, no older than ten, across the street as a round fell near them. Both little bodies hit the wall and collapsed where they'd stood. No way could they have survived such a big explosion so close.

The hatch closed a split second later, and the platoon trucked away from the scene. Lunquist looked between Howard and Benjamin before glaring at Brenner's legs in the hatch.

"That moron is going to get us all killed."

They drove to a safe distance and called EOD. All the complex calculations needed to figure out where those mortars had been shot from were put into play, but none of it would bring those kids back. And not a man in that platoon would lose sleep over it, though some would probably never forget it.

Benjamin had never seen someone die before, and it stuck with him the rest of the night, long after the point where he should have been asleep. He'd known that people would die, and he'd known that he would see it, but for it to finally happen had been strange.

In a way, going to war was the acceptance of death. The plane ride over had been their lowering into the ground, the roar of engines the dirge leading them into Hades. They'd held their wakes in the parties before their departure, the crazed sex with strangers or wives, the alcohol consumed by the bucket. There was no going home, not for them. Warriors embraced their fates the moment they stepped onto enemy lands.

But that didn't mean you wouldn't rather see it happen to someone else. Shit happened, and the only thing to do afterward was move on. In fact, the only conversation about those events after hours and hours of procedure was between him and Garcia the next day.

"Benjamin."

He turned from his walk back from the shower trailer at the FOB, flip-flops on and shower bag in hand, to see Garcia sitting in the chair outside his CHU.

"C'mere a minute."

Benjamin jogged over, nearly sending his flip-flops flying, to stand at parade rest in front of his platoon sergeant.

"Chill, man. Chill." Garcia waved and smiled.

"What's up, Sergeant?" Benjamin dropped out of parade rest but kept his hands and shower bag locked behind his back.

"You heard that horseshit the lieutenant was trying to pull yesterday, yeah?" Garcia asked with a sly smile.

"Roger, Sergeant."

He leaned back in his chair and ran his hand across the stubble on his face. "Next time he does something like that, you stop him. You're the RTO. It's your job to keep an eye on him."

That might have been part of an RTO's job, but disobeying an officer wasn't. "I'm just a specialist, Sergeant. I can't do that."

Garcia patted Benjamin on the arm. "I'll back you up. Just don't let him get anybody killed, hear?"

"Roger, Sergeant."

"Cool, man. Carry on."

Benjamin did just that.

Chapter 4

His life became an ever-widening circle of friends, school, and work. School went well, proof he was doing at least one thing right. He did his homework and reading every night at work and went into school to ace every test and assignment. College was easy after spending so much time in Iraq. Everything was easy.

Or almost everything. For no reason at all, anger wouldn't let him go. Cut off in traffic, a curt word said in passing—any of it was enough to ignite the spark in him.

A young man from math class waved him down on campus as he was heading toward his car. He smiled as he approached, panting from the light jog.

"Hey! Hey, man!" He extended a hand, which Jay shook. "You used to be in the military, right?"

The candid way he inquired about things that were none of his business got under Jay's skin. Civilians always wanted to know what war was like that badly without having the balls to find out for themselves. It wasn't a form of entertainment, and he wasn't going to pretend it was by bragging about his exploits.

"Yeah." Jay adjusted his backpack and did his best to look uninterested.

The kid smiled as the sun shone down on pretty girls and handsome young men all around them, the lifeblood of the campus. "Cool, man. I was thinking about joining the Air Force, and I wanted to ask some questions from someone who's been...you know, *over there*." He leaned in as if they were sharing some great secret.

Those first drops of venom dripped into Jay's gut. The slight burning, slight tingling that let him know anger wasn't far off. The grip around his heart that numbed his fingers and toes. "You'd be better off asking a recruiter."

The kid nodded as if he had all the answers already. Jay looked at the girls. Looked at the math and science building. Anything to occupy his attention besides the stupid man-child in front of him.

When he didn't speak for a few seconds, the young man took it as an invitation. "So what's it like? Going to war? I mean, is it exciting? I hear the pay's pretty good."

Despite what people said, the pay wasn't bad. Not as much as some middle-class jobs, but when you counted all the benefits, it was substantially more than it seemed. "It's hot and shitty." He bit off the words, trying to make it clear he wasn't in the mood.

The boy's smile faltered a little, but he pressed on regardless. "Did you ever kill anyone?" He looked Jay right in the eyes as he asked it.

His blood ran cold. The venom transformed into the fighting itch. The hardening of his heart that told him if he were going to hurt someone, he wouldn't mind, at least not right away.

The scowl on his face must have been a dead giveaway, because the boy stopped smiling. "I don't mean—"

"Hey, man. Go fuck yourself."

The kid stared as if Jay had just slapped him in the face. If he didn't walk away in a moment, he would. He took the hint and did so, but it did nothing to soothe the rage.

Jay walked back to his car in a red haze. He threw his bag into the trunk, slamming it shut only to pound the steering wheel when he sat inside. A young girl watched him before quickly looking away when he made eye contact.

It wasn't shame for the things he'd done; it was anger at the audacity of someone asking that, like asking a man how he preferred to fuck his wife. It was none of their damn business. If they wanted to know, then they would have to make sacrifices on their own. He wouldn't be their goddamned billboard.

While he drove down I-75 sometime later, two cars collided on an overpass above him. The deafening crash in the small, dark space sent him back to Iraq, explosions and gunshots all around. He slammed on the accelerator and shot out from under the overpass, dodging between cars, trying to get out of the blast area of any secondary explosions.

Only there wouldn't be any. Nothing exploded in Tennessee. He laughed at himself the rest of the drive home, calling Allen that night to share both the story and the mirth.

Romanticizing war came easy in retrospect. But fighting, killing, was hard and dirty, and it required those who did it to be equally so.

"Killing another man is not a natural act." It had been so oft repeated he heard it in his sleep. But that wasn't his concern anymore. He'd found freedom from all that bullshit, and he had the DD214[22] to prove it.

Still, he felt a responsibility to those fighting. In an English class one semester, he ran into a Navy veteran with strong feelings on the subject.

"George Bush was a war criminal, end of story. Us taking advantage of that and leveraging it into making the world a better place doesn't change it. It's us making the best of a bad situation, and if we do that, we're still complicit in his crimes."

The other students watched, their mouths glued closed. The teacher leaned against her desk, observing with a wry smile. The exchange had started over how they could best help the people of Iraq. Jay had said they already were. The people needed them now that the occupation was so far along. To leave now would be

---

[22] Paperwork that provides proof of military service and a record of awards received.

homicide for that nation. He wasn't sure if he believed that once the words were out of his mouth, but it was too late to retract them.

The Navy veteran disagreed.

"Furthermore, I hardly think that committing more damage to the mental well-being of their country is going to fix anything. We need to back out and leave them alone."

Jay shook his head before the man even finished. He knew this part of the rhetoric well. "No. No. It's well past that point already. People who say we should pull out now aren't saying it for Iraq; they're saying it for us. They're saying we should leave a whole country to the wolves, and that's bullshit. If you lived there, you wouldn't be so quick to condemn the people helping."

But that wasn't quite right, was it? They hadn't helped as much as they had destroyed. It was hard to be helpful with a gun. The Navy man shrugged and went back to the paper in front of him, ending the discussion, but Jay thought about it the rest of the day and deep into the night.

Did he believe he had helped those people? He certainly had at the time. The words "hearts and minds" had been drilled into their heads so much that it became impossible not to believe them. They had made progress overseas, some attempt to make the world better, at least on their end. He couldn't speak for the politicians that had sent them there.

But that wasn't quite right, either. They had also shot their way across the country. They had screamed at and abused Iraqis when the situation called for it, and even when it didn't. They had slapped and beaten people in front of their own families, occasionally for nothing more than speaking out. They had enforced a curfew on a sovereign nation and sometimes shot people foolish enough to break it.

Could you call yourself a liberator with a closed fist and a totalitarian campaign?

The question plagued him. When he woke the next day, he was no closer to answers.

Chapter 5

Sutter wasn't speaking to him, and that suited Benjamin just fine. For two days now, they'd barely said a word to one another. Sutter had complained he was typing too loud, which any sane person would recognize as stupid.

"Bro." Sutter peeked out from behind the blanket he was using to cover his face. "Do you mind? I'm trying to catch some sleep."

Benjamin ticked away at the keyboard, hard at work on the next Great American Novel. Written at war by a solider, what a selling point that would be. "What are you talking about?"

"Your typing, man. It's too loud."

Benjamin glared at him, pointing at the barely covered window behind his bed. "It's the middle of the day, dude."

Sutter sighed and put on a condescending, *faux* calm expression. "I know that, *Brian-ja-min*, but we have a mission in four hours and I'm tired."

Benjamin kept typing. "Not like you do shit around here anyway."

That spark lit the powder keg. Sutter jumped out of bed, standing there in nothing but his boxers, the fake calm gone. "You know what? Fuck you!"

Benjamin rolled his eyes. "Heard it before. How about you get back in bed before you strain something?"

"Nah, dude. Nah." Sutter took a few steps toward him. "I'm sick of your shit!"

Benjamin finally set the laptop aside. Soldiers were exactly the kind of people who would break something just because someone thought they shouldn't, and replacing a laptop here would be a pain in the ass. "Okay." It was his turn to use the condescending face.

They wasted damn near an hour screaming at one another before Benjamin left to go for a walk. That was the way of things. You couldn't teach a man to kill, to hate, and not expect it to backfire on occasion. He and Sutter would be friends again sooner or later. They had to be. They were family.

But anger was a drug as sure as heroin. It felt good. It fed you. It sustained you when other things wouldn't. Brotherhood, yes, but anger kept you warm in the cold night. These terrorists and Iraqi "patriots" wouldn't keep your friends from going home. Every one of them would die before they touched one of yours. It was no great mystery that so many came back broken. Teach a man to love that drug, and then take it away. Let him go without. Would it become motivation or damnation?

But he wasn't broken, and neither was Sutter. He couldn't imagine any of his people going back and having to see a doctor about the anger issues. The military drilled it again and again. Combat stress. PTSD. Buddy checks. Don't be afraid to speak out if something goes wrong. Still, this was the infantry. If you were hurt, you walked it off. If you couldn't walk it off, you took a profile that lasted only as long as you needed it. They weren't weak, and while never stated, the implication was that PTSD was for the weak, not for them.

They didn't speak for two days. A few episodes of *Deadwood* finally compelled them to make up without a word. Screaming, "Cocksucker!" at the TV together before descending into laughter.

"You and your boyfriend finally make up?" Donner asked on the way to the chow hall with Masters and Simmons.

"You're my only boyfriend, boo." Benjamin grabbed a handful of Donner's ass.

Simmons laughed, and Masters muttered under his breath.

"Don't be jealous, man. Here, let me get a finger in there." He reached for Masters' ass, but Masters dodged out of the way.

"Hey, man. Hey! Hey!"

They laughed and joked all the way to the chow hall, weapons slung over their shoulders and uniforms faded from sweat and sun. They laughed while they ate chicken and rice, looking at the pretty women and the AFN[23] propaganda in equal measure. They laughed on the way back and all the way into the evening, hanging out outside their CHUs while Benjamin and Sutter smoked, the day's heat sending cascades of sweat down their backs.

They joked even as their trigger fingers itched on every patrol. Even as they woke bleary-eyed in the dead of night to explosions, or emergency missions coming down the pipe. There had to be laughter.

Benjamin had watched the news his whole life. The media portrayed the ultra-conservative pro-war crowd as hard-eyed killers with no sense of humor. But their company sniper, Baldomero, had sat in Donner's CHU not three nights before, lip-synching Britney Spears while the mortars laughed along with him. Benjamin had found an old CPR dummy and dressed it up in Nelson's clothes, leaving it on his bunk. Nelson had kept it and named it Combat Steve, taking its head with him everywhere they went as a running joke, talking to it as if it could talk back.

They killed, they were killers, but they were more than that. They were young men in their prime. The greatest of a generation, no different from the soldiers of World War II. Benjamin didn't believe for a moment that the stress hadn't gotten to those soldiers as well.

"Where the fuck is the other radio holder, Olson?" Nelson screamed as they laid out everything from their truck one day.

[23] Armed Forces Network: Military TV network.

The army had endless layouts and checklists. Redundancies to check redundancies that hadn't mattered in the first place. Worries about losing a hundred dollar radio mount while they dropped ten million dollar bombs. Corporal Olson had an excuse for every layout. "I dunno, Sergeant. It was here a minute ago." His high Southern accent sounded childish, even more so when he was nervous. He took a step backward, nearly crushing the GPS.

Nelson pushed him off not a moment too soon, avoiding damage to the equipment. "Watch what the fuck you're doing, stupid!"

Sutter and Benjamin made eye contact as they smoked, neither saying a word. They both knew what was coming next.

Tears glistened in Olson's eyes. "I'll find it." He cried more often than most, though Benjamin suspected everyone was guilty of it on occasion.

Nelson backed down a step, whether from kindness or not wanting to deal with the waterworks, Benjamin couldn't say. "You better fucking find it. Come get me when you do. And you two!" Nelson pointed at Benjamin and Sutter. "Stop being useless and fucking help him!"

Nelson was a good leader and a great guy, but he let the bullshit of the military get to him. Nearly all the mortars did, Miller most of all. But not Donner or Masters. Both sat on the ramp, watching the whole thing unfurl. Neither looked stressed. As soon as Nelson walked away, they joked, going about their business as always. By the time Olson cried in his CHU a few moments later, they had dropped the whole thing and begun to tear the truck apart searching for the missing items. Those kinds of guys could make a career out of this.

Both Benjamin and Sutter had talked about it back in garrison.

"I dunno, man," Sutter said while they stood on the balcony that had been the company smoking area. "I could see myself getting E5[24] and being stationed somewhere down in Texas. Not too far from home, you know?"

Benjamin could see it too. Maybe Ranger School. Maybe Special Forces. If he was going to stay in, it wasn't in the regular army.

But Miller had insisted on an answer to the re-enlistment question as soon as they arrived in country. If the answer had to be forced, then it would always be no.

Searching for meaning every day didn't help. Each mission and patrol bled into the next. Some sense that they were doing something good, something right, hung over it all, but that deeper meaning had yet to reveal itself. It wore him down. It ground at his patience until he snapped at his teammates the same way they snapped at him. And before long, the question of re-enlistment wasn't a question at all.

No, when this year was up, he was getting out. It was a nice fantasy, though. The idea of making a career of this. The thought of

---

[24] Rank is sometimes denoted by a letter and number designation in the military. E is enlisted, and 5 would be the fifth rank up. (Sergeant)

standing proud at the front of a formation until such time as he retired, both his body and ribbon rack heavy with age and overuse.

He lay in bed that night after the layout and the patrol that followed, staring up at the ceiling as he fell asleep. Maybe it was for the best. No way could this crew of killers be topped. Captain Clinger had said it time and again: They were the best. The battalion and brigade commanders called them the tip of the arrowhead in the Arrowhead Brigade. They accomplished their mission every time and to the letter, and all without losing a man.

He rolled over and finally drifted off to sleep. No, there was no military after Charger Company, not for him.

Sunlight poured in as Olson burst into their room in full battle rattle not six hours later. "Get up you two, right now."

Both Sutter and Benjamin rose from their beds. He had his serious face on. No tears this time; he looked…upset? Angry?

"What the fuck is this about?" Sutter asked.

"Get your shit on. Sergeant Henkes is dead."

Benjamin sat up, rubbing the sleep from his eyes. That sort of joke wasn't funny, and Olson should have known better. "That's fucked up, dude."

But Olson's façade didn't crack. "He got killed by an IED this morning. Get your shit on; we need to go cover their patrol."

And then he left. The door closed, and the sunlight vanished. The enormity of it hit Benjamin in the gut like a fist. They'd lost their first man. One of his brothers was dead.

"Fuck." Sutter stared at the wall in front of him.

Benjamin's stomach turned and he thought he would be sick, but there was no outward sign the news had sunk in.

There was no time for that. They had a mission.

Chapter 6

They played games, nothing more. One weekend a month and two weeks a year. In no other profession could a person work so little and be considered competent at their job. How the military thought the National Guard could was beyond him.

The military trained killers. If not killers, then people who supported them. All the talk of patriotism went out the window when the bullets started to fly. It wasn't about the mission, the country, or the people back home. It was about your brothers on your left and right, about killing the other fuckers before they got you and yours. Blood, blood, blood made the green grass grow, not flowery words about changing the world. He and his brothers had trained from day one to make that grass grow, and they were good at it. There had been no better-trained killers in the world. These people weren't the same. They weren't his brothers.

The staff sergeant at the hood of the car stood there holding out the mock rifle as if any moment the woods of Tennessee would burst to life with jihadists. He scanned back and forth, taking in the old drill hall and field full of cows next to it.

The bovines mooed for the third time in ten minutes, and once again, Jay had to stifle a laugh.

"Okay, lookie here." The overweight squad leaders ACUs practically screamed as he dug the muzzle of his fake rifle into the dirt and leaned on it. "First squad, I want you to go check out that building. Second squad, I want you to set up a TCP at that intersection."

One of the other specialists, a man not much younger than Jay but lacking the combat patch, raised his hand. "Who's going to watch the car?"

The "scenario" they'd set up was simple. A car stalled at an intersection with two Iraqis inside acting suspicious. Iraqis always looked shifty to Americans, but not as shifty as this training exercise. Nobody took the drill seriously. Why bother? The last deployment the Tennessee National Guard had been on involved gate guard and convoy security for a year. A bunch of farmers and blue collar workers going overseas to pretend to be soldiers.

Jay squeezed his rifle tighter at the thought of it. They bragged about their exploits as if they'd done anything even approaching the things Jay and his brothers had done overseas. They hadn't. A bunch of POG[25] cav[26] chumps pretending they were infantry.

---

[25] Person Other than Grunt
[26] Cavalry

The squad leader seemed to consider the question for a moment before he turned to Jay. "He can."

A mortar attached to a cav unit, and somehow he still got treated like the farm cousin in the big city. Or vice versa, as the case may be. "I'm the terp."

"So?" The squad leader furrowed his brow and puffed his chest out, clearly annoyed at the lack of respect and questioning of his orders. Little surprise that a guy like this came to a place like that to exert what little authority he had.

Jay bit back the scathing response on his tongue and instead responded as level as he could. "So, a terp isn't a soldier. You don't leave them to guard shit, and you certainly don't leave them unattended… Sergeant."

The staff sergeant scoffed. "You'd be surprised what you do in Iraq."

For an instant, the whole world was soaked in red. Jay's anger clawed up his throat and out his mouth in a condescending laugh. Every fiber of his being wanted to grab that fake-ass rifle from the slob and slap him upside the head with it. Kick him in the knee and watch him crumble to the ground. Show him how much that Stetson and those spurs were really doing for him.

Then sanity reasserted itself, and Benjamin blinked several times before he remembered who he was and where he stood.

Now both squads stared at him. He glared at the squad leader, and the man looked taken aback by it.

"I've been stateside less than a year. The only thing surprising is that your whole squad made it back alive."

An angry grumble escaped the crowd, but Jay was already walking away before the staff sergeant could respond. He half expected to be told to stop. To have some quivering mass of Tennessean come barreling at him, demanding the respect his rank and fifteen years in the army were due.

It didn't happen. The man let him walk away, and Jay resented him all the more. In fact, nobody said a word to him about it. Not that drill, or the next one, or the next one, but it didn't make him any friends.

"You could probably put a little more effort into that," one of the other mortars said as they sat around their office two drills later. "Not like you're giving off that 'come hang out with me' vibe."

Jay hadn't so much as bothered to remember the man's name. He glanced down at the name tag on his chest before responding, not giving a damn one way or the other if he noticed. "Fuck'em, right? I'm just running out the clock here."

His not-quite-teammate shrugged. "Tough way to go through life, man."

But it wasn't life. It was *this* life. This half-military, half-civilian hybrid of a thing. Not out, but by no means in any longer. Not fitting in with the people who wore the same outfit, but still part of the greater whole.

As he drove home that weekend, the thought kept turning over and over in his mind. *Out.* When he'd been active duty, the word held an almost sacred space in his mind. It hadn't before Iraq, but by the time they'd been there six months, it certainly had. He envied the dedication of the guys still in, but he wouldn't make a career out of it.

At least, not intentionally. Now that he'd left, he wasn't sure he hadn't. He'd stopped wearing the uniform daily, but that mentality, those thoughts, never left him. The feeling that he was still a part of something greater, that it all had some meaning or purpose. That by being part of it, he had some larger meaning. Even beyond helping people with medicine.

By the time the next drill rolled around, he knew what he needed to do.

Sergeant First Class Murdock sat behind the desk in the training room going over paperwork when Jay knocked on the door and let himself in.

"Hey, Benjamin." He set the papers down and leaned back in his chair. "What can I do for you?"

Jay snapped to parade rest, that part of him reflexive even if the rest of the unit had given up on propriety. "Hey, Sergeant. I was wondering who a guy had to talk to about switching his MOS[27] to medic?"

---

[27] Military Occupational Specialty

And why not? Being infantry had made him the man he was, but something needed to change, and being a medic was just as good. It would help him get into medical school, and with a little luck give him the kick in the ass he needed to find the thrill, the hunger that had made life after the army seem so tantalizing.

"I'm your man, but I'm afraid we don't have any medic slots at this unit."

"What about other units?"

Murdock shook his head. "Not unless you're keen on travelling across the state every month and two weeks a year. But let me make a few phone calls. I'm sure we can figure something out."

Benjamin left the office and spent the rest of the day doing the same thing he'd been doing with these people for months now. Sitting around the office, staring at one another. Playing cards. Exchanging stories about Iraq and the glory days now long behind all of them.

But not him. He wouldn't let the most important thing he'd even done be over before twenty-two.

Murdock called him into the office before everyone left for home that night. "Nothing here in the state military, but there's some reserve slots if you're okay with going back to federal."

Jay smiled. Anything was better than living in limbo.

Chapter 7

A fluke, nothing more. An IED had gone off just a little ways up the road and a piece of shrapnel had pierced his eye protection, killing him. The details floated around the company, impossible to escape.

It was one thing to be blown up or shot at; it was another to stand beside all your friends in the MWR tent while your chain of command spoke about what an upstanding guy your teammate had been. Jumping at the twenty-one-gun salute was different when they were honoring one of yours.

But mourning would come later.

Tension rose across the next days and patrols. Everyone was on edge. Donner and Masters barely spoke. Simmons, Sutter, and Benjamin bickered constantly. Miller and Nelson kept a straight face through the whole affair, but they had to. It was their job to be stoic during tragedy. They had an example to set.

Things in the back of the truck with the riflemen weren't much better.

"We should glass this whole fucking country," Quinn said as they sat on outer cordon for a company operation in town. "A few fucking nukes would show these goddamn savages what was up."

Brenner said nothing, but it was a poignant nothing. Benjamin could almost hear the objections. *Hearts and minds,* he would say. *They're people, just like you. If we're going to succeed in this war, we need to win them over. You can't lose faith in the goodness of others. That's what separates us from them.* But he didn't say it. He knew better. Blood stained the water now, and the sharks were on the hunt. Nobody wanted to hear about allying with the people who'd murdered Henkes. Iraqis weren't bad, but sometimes it felt that for every decent man in Iraq, there was a terrorist. Bomb makers who killed your friends as they drove down the road. Snipers who shot at you every time you put your head out of a hatch. And you could never tell one from the other. The little old ladies walking down the street were just as apt to have bombs strapped to their chests as everyone else.

Brenner and the chain of command were right, of course. The only way to win this war was by having the people realize the Americans had their best interest at heart. But wasn't that disingenuous? Hadn't they invaded this country guns and missiles blazing? Hadn't they destroyed infrastructure, taken over government, and enforced their own laws and customs onto a nation?

Howard kicked Benjamin in the leg. "What's your problem?"

"Nothing." Benjamin shifted his rifle from one knee to the other.

"Your cooch bleeding?" Rosenberg asked from the gunner's seat.

"Just a bit."

The others chuckled. Quinn lit a cigarette, though it was against the rules. Against *some* rules, anyway, but not his. In his house, his word was law. "I'm telling you guys. I've been here often enough to know that they're fucking animals. These shit birds have been murdering each other since they had to do it with rocks. That isn't going to stop in our lifetime. If I were in charge of this shit show, we'd have blasted them straight to fucking Allah by now."

Howard kicked Benjamin in the leg again. "I don't think Benjamin would do it. He's a hadji lover, Sergeant. He wants to buy them flowers and meet their dads."

Quinn fell silent for a moment. A smile spread across his lips. Mischievous, not malicious, though the two often looked the same on soldiers. "That true, Benjamin? That what you thinkin' 'bout over there?"

Easier to deflect than to answer. "No, Sergeant. I was thinking about fucking Howard's sister."

With the exception of Brenner, all the men in the Stryker cracked up laughing.

"Well you better change MOSs, then," Lunquist said. "I hear she likes MPs[28]."

"I hear she likes anything you throw her way." Quinn took another drag off his smoke.

Howard laughed along with them but cast his gaze to the floor. "That's fucked up, guys."

Through the night and into early morning, the mood shifted from joking to serious to maudlin and back again. Hard words were expected from hard men, though Benjamin wondered if Quinn would really do it given the chance. Many people had said they'd wipe out this country and everyone in it. That they'd wipe Islam off the face of the Earth if they could. Every time Benjamin heard it, he wondered if it was true. They were his family, his brothers. He knew them to be killers. He was one himself. But could they wipe out an entire culture?

Would they?

Things continued as normal, as if a man wasn't dead and a daughter wasn't without a father. As he sat in the Stryker with Brenner and Jared on QRF, his thoughts kept coming back to that. How many fatherless daughters had they created? How many orphans ran the streets of Iraq because of coalition forces?

---

[28] Military Police

Hundreds? Thousands? Even more? It was easy to pretend American lives had more value. They were worth more to each other; that was understood. He'd eat all the children in this country before he let his brothers come to harm. But how much did that matter in the scheme of things?

"It's okay to talk about it if something's bothering you." Brenner had stayed in the Stryker to read while Benjamin and Jared were on radio duty, but his book now sat open and unnoticed in his lap.

"I'm fine, Sir."

Brenner looked off into the middle distance. "Did you know Sergeant Henkes?"

"Not really." Benjamin didn't make eye contact, instead staring into the filthy parking lot as the call to prayer rang out across the city. Ripples of heat rose from the black top.

That truth made the whole thing harder to stomach. He hadn't known Henkes at all, and now he was gone. Much like an uncle you didn't spend enough time with before they suddenly passed, it left an absence. A wondering about what you might have missed in getting to know them. Others spoke of what a good person he was, what a capable leader. They hadn't *always* said that, of course. There had been just as much bitching as there had been respect before his death. But those things ceased to matter once they were gone, didn't they?

"I didn't know him too well, either. He struck me as a very responsible and caring man. I know his lieutenant spoke highly of him."

Benjamin didn't want to talk about it. He wrote letters every day to a woman back home, Tiffany, and the one he'd sent her the night Henkes died had been a real tearjerker. He'd regretted it as soon as he sent it. She didn't need to know about things going on here. Better to lie to her and let her think he was safe than to have her worry. Still, he'd already purged those demons, and he didn't feel like revisiting them.

He finally faced Brenner. "I don't think that—"

The first mortar landed in the parking lot with a deafening *boom.*

Benjamin jumped in his seat, his finger instinctively finding the trigger guard of his rifle. "Shit!"

Boom!

The next mortar crashed against the side of the building, raining concrete onto the trailers in the parking lot. Benjamin watched, mouth agape, as it struck. He'd never seen one land that close before.

Boom!

Another collided with the ground, kicking up a mushroom cloud of dust. It missed all the vehicles but sent chunks of dirt soaring through the air.

The few Iraqi soldiers near their vehicles outside made a break for the building. Nobody had been hit that he could see, but he wasn't going to run out there to find out.

Brenner had other ideas. "I need to go check on everyone inside!"

Benjamin's hand was already on the toggle that would raise the ramp and guarantee their safety. Even a direct hit from a mortar wouldn't pierce a Stryker as far as he knew.

"Just stay here, Sir. They're inside, they're safe."

But Brenner had already bolted from the Stryker, running the thirty feet across the open as more mortars fell.

Benjamin raised the ramp without a second thought, watching the mortars land with their ear-splitting booms until the portal sealed and he felt more than heard them. Brenner reached the building safely just before it snapped closed.

Jared, who had been quietly sitting farther back in the Stryker, had already shut the hatches on the roof. They made eye contact until the explosions stopped. When only the ringing in their ears remained, they laughed, as if mortars falling and their lieutenant running off into the thick of it was the funniest thing they'd ever seen.

"That guy," Jared said in his thick Mississippi accent, "is a fucking idiot."

Everyone inside had been fine too. The rest of QRF passed without incident, and they returned to the FOB. In no time at all, they were back out on the road, picking up a bomb maker who'd blown himself and his house to shit from an Iraqi hospital and taking him to the one on the neighboring FOB.

They loaded up and rode out, conversations going on as normal. But just beneath it was tension. Had this man killed Henkes? Had one of his bombs murdered their brother?

The ride to the hospital took no time at all. Quinn stood in his hatch, Lunquist and Brenner from the rear ones. Soon, they found the entrance and received an overweight Iraqi man in a wheelchair.

"Oh. Mister. Hurts. Hurts."

The man from the ER rolled him out and left him at the open ramp to the Stryker without a word. Who could blame the guy for being disinterested? People who weren't involved with the war were in a damned if you do and damned if you don't situation if there had ever been one.

"So this is the piece of shit bomb maker, huh?" Quinn kicked the chair, raising another wave of moans. "Get him in the Stryker. Put something down. I don't want hadji blood all over my truck."

They placed a poncho over the seat and eased the man onto the ramp and into the vehicle. A few moments later, they left for the FOB. Every bump brought a cry from the man.

"Does that hurt?" Howard asked. "Good, you piece of shit. I hope it does."

If the man understood English, he pretended not to. The others jeered, and to his surprise, Benjamin found himself joining in.

"Piece of shit."

"Wish you'd killed yourself too. There wouldn't be seventy-seven virgins waiting for your fat ass."

"Terrorist fuck."

Before long, they arrived. The ramp dropped, and Olson got out of his truck to run into the building and grab a wheelchair. He and Benjamin eased the man out of the Stryker and into it while the others glared. The insults stopped in "polite" company. Soldiers whose jobs didn't involve violence watched as dirty, hard men removed the bleeding one from the Stryker. The job of taking him inside fell to Benjamin and Olson as Miller, Brenner, and Garcia went ahead to deal with the bullshit and the paperwork.

"Mister. It hurts. Ohhh, it hurts."

Bile rose in his throat as Benjamin pushed the man toward the door. *Murderer. Terrorist.* All the labels came to mind, and for one insane second, he almost dumped him from his chair. Why not? Henkes' daughter would suffer much worse in all the years she'd miss without her father. She'd have to deal with stares and pity when her schoolmates and later co-workers found out her father had died in the war. What was a little more pain for this asshole?

"Are you all right, Benjamin?" Olson asked.

He hadn't realized he'd stopped at the small curb that marked the edge of the street while he considered his options.

"Mister, it hurts."

He didn't dump the man. Instead, he muttered, "Shut the fuck up, hadji," under his breath as Olson helped him navigate the bump and take the man into the nice, air-conditioned building. Two medics wheeled him away. He'd receive medical treatment. Food. Water. Later on, when he was better, he'd be sent to a prison for a few years at worst. Even that much probably wouldn't happen. Within a year or two, this man would walk the streets of his hometown again as if nothing had ever happened. Maybe the locals would hail him as a hero.

Henkes would never get that.

As soon as they passed him off, Benjamin left, pulling out his cigarettes before he'd even reached the Stryker and lighting one. The others still grumbled, still eyed the door as if they were thinking the very thing he had been. Maybe worse.

Masters summed up their feelings nicely. "Fuck that mother fucker."

Just a few years ago, Benjamin would never have imagined dumping a wounded man into the street and smiling while he bled. But things were different now. He was different.

They all were.

Chapter 8

"It's a different Iraq, man," Masters said over the phone as Jay lounged on the couch in his apartment.

He'd always kept up with his Charger brothers. Now that they were back overseas, it felt more important than ever. Many had gotten out of the military or moved on to other things, but enough of the '06-'07 Chargers were left for him to care.

"Yeah?" Jay looked around at the life he'd made. It wasn't much. The same apartment since he left the service. Working the same security job, though halfway through school to be a surgical technologist. No real progress. Nothing to show what he'd done since his deployment. It still felt like borrowed time, as though he should have never left Iraq. He shifted in his seat, uncomfortable with the thought.

"Hell yeah. We have to stop at fucking red lights now. Can you believe that?"

During the troop surge[29], it had been military first and civilians a distant second. They blew through lights, ran over cars when needed, sideswiped vehicles damn near everywhere they went, and shot at anyone stupid enough to get close. A sign had hung from the back of the Stryker that said in both English and Arabic anyone driving too close could be killed. They'd been all business.

"It's real different. We hardly ever get shot at. Almost no IEDs. Most of what we do is training and hanging out. It's almost like being in garrison. It's kinda bullshit."

"War not fun enough for you anymore?" Benjamin smiled.

"Shit. I already got to do that crap. I'll take this any day. They ain't ready for us to go, though. I think if we left now, this place would fall to shit."

Bush had planned the troop withdraw. It was coming soon under Obama. Many on both sides screamed about it. One insisted it was too soon, the region still too unstable to allow Iraq to support itself. The other didn't care. Enough American time and blood had been spent there. A recession had hit, after all. The US had to look after its own before it could police the world.

Neither side gave a damn about the Iraqis who'd spent their lives trying to make the best of a bad situation. Even Iraqi politicians seemed less than concerned about the state of things than they should have been, demanding the Americans leave, though few could blame them for that. But Jay cared. Certainly, some veterans

---

[29] Massive influx of US forces to Iraq in 2007.

came back callous toward the people of the Middle East, but many returned with a keen interest in seeing them better off than they had been. After investing so much sweat and tears, there was no other option.

"So you want to keep going back over there?"

Masters laughed. "Fuck, man. Easy money."

It hadn't been so easy last time, but conversations like those always put Jay's nerves on edge. "All right. I'm going to let you go, man." He felt bad for saying it, but he'd gotten out for a reason. "You be safe, okay?"

"I will," Masters said. They both hung up.

Thinking about everything and nothing, Jay stared at the water stain on the wall coming from inside the ceiling. He wondered what the world looked like to his friends who'd stayed in, and whether or not they'd made more of their lives than he had.

It really was a shithole apartment. Not bad for a college student, but he should be more by now. Was it entitlement? Had he bought too freely into the propaganda of soldiers deserving more? Had it really been two years? Time, the big machine, churning the Earth as you passed. During easy times, you didn't see it. But when it became hard and you tried to stop, it was right there behind you, pushing you along no matter how slow it first seemed.

He'd worked security at a Tennessee welcome center a year before, a little rest stop at the Tennessee-Georgia border on I-75. Covered in trees and near the Tennessee River, it was an idyllic spot where you'd walk your dog and smile, thinking of the vacation you were traveling to or from.

An old man and his wife had stopped there on their way to Florida to visit grandchildren. The man had picked Jay out as a veteran right away.

"You a military guy, huh?" He walked up and leaned against the counter. His hair was white and his skin pale, but his eyes shone with intelligence. He smiled as he spoke.

"Used to be. Now I'm mostly just this." Jay gestured at the desk where he sat and the vending machines on the far wall.

The old man rapped his knuckles on the counter. "Nothing wrong with that. I did a lot of jobs when I first got out too."

"You're a vet?"

The old man smiled. "Korean War! Hard to believe it was so long ago now." He looked old, but not the seventy or eighty years he must be.

"No kidding!" Jay stuck out a hand. "Well thanks for serving, sir. Always good to meet a fellow veteran."

The old man shook his hand and laughed. "Long time ago. Long time." For a moment, his eyes found the space between things as he stared into memories. What was he seeing then? Himself in the mirror when he'd been young and strong? The faces of his friends who hadn't made it home? Just as quickly, he returned to the conversation. "Met my wife right after that." He gestured at the women's restroom. "We travelled. Went to college together. Got a job and had a whole bunch of kids. You have a wife?"

Jay held up his hand without a ring on it. There were many dates, but no Mrs. Benjamin.

"Ah! That's all right. It's good to share your life with someone, but you're young. Someone'll find you eventually."

"The world's my oyster, sir." Nothing could stop Jay then. He'd finished a semester. He was starting surgical technologist school soon. Things were looking up.

The man threw his head back and laughed as his wife walked out of the bathroom. A rich, hearty sound, something well used. "That's the spirit, son!" He reached across the desk and shook Jay's hand again. "Never let that feeling go."

Jay had gone home that night, then the next, then the next. Soon, nearly a year had passed, and everything felt different. Things weren't bad, per se, but the world didn't feel like an oyster anymore. His friends—Nelson, Masters, and the others still in Charlie Company—were back overseas. That belonging, knowing you were part of a bigger picture, was gone.

Maybe every cog felt this way once removed from the larger machine. They hadn't changed the world as they'd been promised, but maybe Masters and the others still could. That would still make him part of it. He'd contributed.

He made himself dinner in his cramped, dirty kitchen and sat down to dine alone. Would that Korean War vet still give him the same advice? Was he even still alive? Jay's Great Uncle Cliff had been on the beaches of Normandy, and he'd said the same thing. After the war he saw the world, became a part of it. Jay didn't feel like a part of it. He sat on the sidelines, watching.

The dishes went into the sink, same as every night. He went to bed at eleven, same as every night. The next day he repeated the same routine.

Not upset, just frustrated with the state of things. Angry that so little had come of so much. No different from any other day.

Chapter 9

"Listen the fuck up!" Nelson yelled.

Masters and Simmons poked and punched each other on the ramp while Miller debriefed them. They'd come back from a six-hour patrol, and the company had a mission in three hours. Garcia mandated that nobody bother leaving. He didn't feel up to hunting people down.

It translated to, "Sit your ass on the truck and waste your free time."

At the senior NCO[30] meeting that had followed, more information had been put out, big news. Not so big that two privates would stop playing grab-ass, but still big.

"We're going to Baghdad."

That shut everyone down, and suddenly all eyes were on Miller.

---

30 Non-Commissioned Officer

"I know there's been a lot of shit going through the rumor mill, but it's official. We're going to be handing off with the unit that just got here and heading up there in early December."

"Jesus, that's in like three weeks," Benjamin said.

"Thanks for the calendar update, Sherlock," Nelson replied.

Mosul was hardly safe, but it was better than when they'd arrived. IEDs blew up on Strykers less often, raids were successful, and overall, violence marred their sector less than it had in years. But the heart of the war beat in Baghdad. Everyone talked about what was going on down there. How crazy it was. How this big troop surge starting soon was directed there to take back out-of-control areas of the city. Rumors flew about the militants and the things they were doing. Enough bombs to wipe out BIAP[31]. Ties with the Iranian military. Eastern European mercenaries hired to snipe Americans. Even the craziest rumors seemed plausible.

"Be ready to pack all your shit and go," Miller said.

After more routine notes, the meeting ended and the lower enlisted—mortars, riflemen, and tankers—stood around outside the trucks.

"Guess that means were spending Christmas in Baghdad, boys." Sutter took a drag off his smoke.

"That's in December." Donner mocked Benjamin with a sidelong glance.

---

[31] Baghdad International Airport

Iraq was Iraq. This would drag on no matter where they were. He eyed the T-barriers that separated their living compound and company CP from the small motor pool they'd made outside. The edge of the FOB stood only a few hundred meters away. Nothing more than a berm, some razor wire, and a single guard tower between them and the world outside. Nothing about the war had been as he'd expected.

The war.

It felt bizarre to think of it like that. War seemed so total in the movies he'd watched and stories he'd heard. This was…well, certainly dangerous. Hardly a day went by where *somebody* wasn't shot at or hit with an IED, even if attacks were fewer than they had been months ago. It wasn't always fourth platoon but often enough. But a sense of progress had yet to show itself. There was no great enemy to fight. More often than not, what they found on these raids were scared people. Sometimes more. A man trying to explain why he had a hundred pounds of explosive in his house. A woman at a TCP who wasn't a woman at all but a high-value target dressed as one to try to slip by.

Maybe tonight would be different. Maybe they'd face off against something other than a yokel. Maybe when they got to Baghdad, they would find a real enemy to fight and a real war to win.

Stranger things had happened.

Transitions always invited danger, and this was no different. The Iraqi soldiers at Resolve escorted a man from the small prison building in the back. The man, the prisoner, had taken advantage of the chaos by raping and killing the daughter of one of the men at the outpost, murdering his entire family afterward. All because he worked with Americans.

Jared spat on the ground while Howard and Benjamin smoked, their eyes on the man, the monster, behind them. Masters muttered under his breath, "Fucking pig."

Indeed he was. He also limped as he walked, a fact that didn't go unnoticed by Brenner.

"Why is he walking funny?" he asked the terp. The man translated, and he and the Iraqis spoke back and forth for a moment.

"They raped him."

Everyone froze.

"What?" asked Brenner.

"They raped him with AK and broomstick for what he did to his daughter."

Blood stained the back of the man's pants. The Iraqis handed him over to Benjamin and Jared. "Uh… Sir?"

Quinn, who stood nearby on the ramp, laughed. "You got what you deserved, didn't you, fucker?" He turned to the lieutenant. "I don't want any hadji ass blood on my Stryker, Sir."

Brenner ignored him. "And what are we supposed to tell the soldiers at the prison?" He eyed the Iraqis. "How do we explain a war crime?"

They all argued for a moment, voices raised and tempers high. Benjamin tuned them out. Brenner tried to be righteous and angry about it, but his heart wasn't in it. The man had raped and murdered a child. He'd killed a little girl for her father's political beliefs.

Sometimes he wished he could hate them. It was impossible. For every bastard who killed families, men like the Iraqi guys at Resolve were fighting right there beside Charger Company. For every civilian who aided the enemy, a hundred weren't involved and a few dozen were helping the Americans.

In a way, he pitied them. If he never set foot in this country again, it would be too soon, but he didn't hate them. They were trying to get by, the same as everyone else. Much like back home, there were those who took advantage of chaos and ran amok. In the United States, a system was in place to protect people from it. As much as people bitched about it in the US, Iraq had no such system. That anarchy so many people claimed to crave back home wasn't nearly as pretty as they imagined. Few who fancied they would be leaders of the pack in a world gone mad would have their wishes granted. Sheep always outnumbered wolves.

Sutter laid down a towel while Brenner and Garcia worked out the last details. Benjamin and Jared helped the man up the ramp. For the second time in a month, Benjamin wished he could push a wounded man to the ground and get away with it. He probably could have.

"Owww. Owww!" the man cried.

"Shut the fuck up and walk." Benjamin half-lifted the man onto the ramp.

Quinn sneered before entering the vehicle and getting back in his hatch. "Make sure he sits on that towel. If there's any blood on the seat, you two are cleaning it up."

They loaded up and drove back to the FOB, making for the prison where the other captives were kept. As expected, the soldiers there made a huge stink, demanding to know why he was wounded. Garcia, Brenner, and Miller stayed back to deal with that while Benjamin and the others waited at the trucks.

Sutter, Donner, and Ski sat on the back ramp while Benjamin stared at the puddles of water everywhere. It had rained briefly earlier that day, but most of the moisture evaporated before they had a chance to enjoy it. The cold season had settled in, but the sun still scorched the land.

Nothing enhanced the beauty of this place. Peeling paint on the walls. Cracks in the stone. Pavement so damaged that it was as much dirt as road. Why anyone would want to live here was beyond him.

"That's some straight street justice they gave that guy," Sutter said to nobody in particular.

"Yup," Donner replied.

Masters shrugged. "They should have just killed his ass."

Most of the guys nodded in agreement. Such was the way of life in the infantry.

But there was no time to get comfortable. No chance to sit and ponder the deeper meanings of war and what it made of men. The army had an old adage; hurry up and wait. Truer words had never been spoken. They waited an hour only to hurry back to the FOB to continue packing. When they weren't putting their things away to be carted to BIAP, they helped supply with the endless duties. They spent the last of their time helping the new unit acclimate to the battle space, a momentous task in itself.

To say the unit was incompetent was an understatement. Their codename on the radio was "Demon", which made perfect sense. Benjamin imagined they'd come from hell to torture Charlie Company.

Their RTO conferred with Benjamin while he checked the radios before a mission one night. "You sure mine are okay?" the PFC asked him.

Benjamin had a laundry list of things to do before they could leave the wire. They'd been packing supply until twenty minutes before, and now he had to check all the radios and his gear in only fifteen minutes. "You filled it, right?"

"Yeah."

"And you did a radio check?"

He didn't make eye contact. "Yeah."

"All right then, man. Move out. I'm working."

79

The private ran off without another word. They left the wire right on time. But even before they'd made it to the street, things went south.

"Hey!" Quinn yelled over the headset.

Benjamin sat in the Stryker and couldn't see what was happening outside.

"Tell that fucking moron in the back of their Humvee to turn his fucking weapon to safe. I can see that shit from here."

Benjamin grabbed the handset out of the cradle behind him. "Hellfire 6, this is Green 6 Romeo. Over."

He waited a moment, but no one responded.

"Hellfire 6, this is Green 6 Romeo. Over."

"Holy shit," said Quinn. "Are they rolling without comms? Are you fucking kidding me?"

They'd broken two cardinal rules of the battlespace in no time flat. The men in the Stryker made fun of them all the way to Resolve.

"Leaving this place in some real capable hands." Howard shook his head.

Lunquist shrugged. "We're getting that big green weenie. Why shouldn't the Iraqis?"

When they arrived, Quinn kicked Benjamin in the back as he stepped out of the truck, nearly sending him flying off the ramp. "Go unfuck their radios. If I have to talk to that RTO, I'm going to choke somebody."

Benjamin spent the next forty minutes teaching their RTO how to properly fill and check a radio. How he'd managed to get his position with his stunning lack of professionalism was beyond him.

But that wasn't the end of their ineptitude, or their bad luck. A graveyard sat on a hill at the center of Mosul. Huge and old, it stretched in every direction with taller buildings surrounding it.

The day after the radio incident, they stood in the graveyard with half of Demon Company, running tests on Iraqis to see if they'd handled any explosives. They put a substance on the Iraqis' hands; if it turned colors, they were screwed. Some had already been caught with weapons that didn't belong to them. It never ceased to amaze Benjamin how often Iraqis happened to "find" things that weren't theirs.

"Jesus fucking Christ," Quinn said over the headset as they pulled up. "Someone's going to get shot out here."

The ramp dropped a moment later. Benjamin strapped on his radio and followed Brenner into the winter cold. Soldiers of all ranks stood around bullshitting with one another or staring at the detainees. Nobody looked outward at the hundreds of gaping windows and rooftops all around, any one of which could be hiding a sniper.

"Fuck," Benjamin muttered under his breath. "Can we go home, Sir?"

"Keep it professional," Brenner replied softly.

They wove between graves to meet the captain, Demon 6. The Iraqis looked shifty, but they always did when zip-tied. Who wouldn't? More concerning was all the open space. When he took a knee and began pulling security outward, one of the E5s standing nearby nudged his buddy and pointed at Benjamin as if he were out of his mind.

Brenner spoke to Demon 6, and Benjamin tuned them out. These people were supposed to be their peers? Captain Clinger had said time and again that Charlie Company was the best unit he'd ever worked with. The colonels on both the battalion and brigade level had mirrored that sentiment. He'd always assumed every CO[32] said that to his troops, even if they were a shit sandwich. It came with the rank, no different from privates fucking up or specialists goofing off. But these people had come to replace them. A couple of sergeants snickering as an infantryman did his job.

The snickering ended at the crack of the first round.

A bullet hit *something* with a thud, though in the tense seconds after the shot rang out it was hard to tell what. With the wide-open space and buildings surrounding them, it was impossible to place where it had come from. Soldiers everywhere dove for the ground.

"He's hit!" someone yelled.

Benjamin looked over his shoulder, nearly sick at the thought of Brenner being hurt. He wasn't, but the shot had winged Demon 6 in the arm.

Benjamin keyed the radio. "Shots fired. Shots fired."

---

[32] Commanding Officer

Quinn responded in an instant. "Yeah. We heard. We'll pull closer."

Demon 6 barked orders and fled to his Humvee as they released the detainees. Brenner pulled Benjamin to the ground with him.

"Stay low!"

Benjamin rolled his eyes. His nerves sang and battle high had kicked in, but they were on a hill with buildings all around. If a sniper wanted to kill him, he was going to be killed.

They jogged back to the vehicle as it pulled closer, the remainder of Demon Company returning to their Humvees as well. The ramp dropped as Benjamin watched their surroundings, expecting the next shot to come for him or Brenner. It never did. A moment later, the ramp rose and sealed them in the Stryker, as safe as anyone could be in a war zone.

"That was fucking stupid!" Benjamin ripped the radio pack off his back. "Those people suck at their goddamn jobs." His hands shook as he set the pack aside and cradled his rifle.

"Yup," Howard said. "Those fucking morons are going to lose everything we've done here."

They would. Everything they'd worked for over the last few months would go up in smoke.

Benjamin's hands kept shaking. He wasn't afraid, just jolted. People shot at them often enough, but it was seldom that close. Demon 6 had been only a few feet away. Adrenaline coursed through him. Already the events of the last ten minutes blurred. He grabbed a bottle of water from the rack behind him and downed the whole thing in one gulp, the liquid sloshing painfully in his stomach.

"You all right, man?" Rosenberg asked.

Nothing was all right. Not the stress of moving. Doing missions. Being in a country where a good chunk of the population wanted him and his friends dead grated his nerves. Watching the new folks in town and knowing the meager progress they'd made would soon vanish was too much.

"I'm fine."

Chapter 10

"Platoon!" He gave the prepatory command, and the gaggle of teenagers snapped to attention before the next command left his mouth. "Attention! Left face! Forward, march!"

The company of new medics marched off the company CP and to the small dirt diamond a few hundred meters up the road, dropping their bags and waiting for further commands, though not without a fair amount of grab-assing. They weren't soldiers yet, just wet-behind-the-ears privates and college specialists who thought they were. Benjamin eyed them all without a movement, taking it in. He'd been the same way when he was fresh out of basic training.

The reserve unit up the road from his apartment hadn't been so different from the National Guard, but they had a better sense of who they were. No tired old men grumbling about the days when they'd been real soldiers, the reservists knew their role. People bridging that gap between real soldier and civilian, running out the clock to make sure they wouldn't be called back up and sent overseas.

A few high-speed privates had dreams of making a career of the reserves, but for the most part, the stories people told on drill weekend were of their current lives, not the glory days long since over.

Master Sergeant Wells, the training room NCO, had finally approached him about school months after he'd finished the transition. She shoved paperwork into his arms when she found him telling stories of his many dates in one of the smaller offices of the oversized and underused drill hall. "You're going to medic school next month. Do you need a plane ticket?"

That month flew by, and the week of in-processing as well. In the blink of an eye, he stood before a platoon of soldiers once more. The thrill of being back in uniform made him light giddy. He'd been picked as the only veteran in the platoon to be the platoon guide, and even a few years out of practice barking orders, it returned as natural as breathing. He had Miller to thank for that.

"You always train for two positions above where you're at. You have to be able to do your boss' job in case something goes wrong."

It wasn't just sound advice for the military but good advice for life. As obnoxious as the guy had been, he hadn't been an idiot.

Benjamin smiled, thinking of those days gone by, as he handed the platoon off to the E7[33] in charge.

"What the fuck are you grinning at?" The platoon sergeant didn't appear angry when he asked, only curious.

---

[33] Rank is sometimes denoted by a letter and number designation in the military. E is enlisted, and 7 would be the seventh rank up. (Sergeant First Class)

The rough dialogue only widened Benjamin's smile. "Just looking forward to some of that good army training, Sergeant."

The platoon sergeant shook his head. "Get the fuck back in formation, clown."

They laid out gear, the privates at first doing so with discipline before descending into grab-ass games. Jay barked and hounded them, but there wasn't much point. Privates had no clue what discipline meant, let alone how to display it for more than an hour at a time.

By the end of that Friday night, he'd set up in his new bunk. Another veteran, Diego, occupied the one below him. He'd been cook in a past life, but they were all medics now, or would be soon. Jay rolled over and ducked his head over the side of his bunk.

"What's up, man?"

Diego removed his head from the book he sat reading. "What's good, dude?"

Jay grinned at him. "We have until Monday to do whatever the fuck we like. What do you say we go get shit-faced and get the fuck off base for a bit?"

Diego furrowed his brows at his new bunk mate. It was always so hard to tell with military guys. Like new dogs discovering one another in the street, there were lowered tails and flattened ears. Maybe a little baring of teeth and a growl before they sniffed one another and moved on or butted heads.

But Diego laughed. "You're buying, right?"

Two hours later, they and a few others drove off Fort Sam and out into San Antonio. A far cry from the mountains of Tennessee, but not so different than home. Tall buildings, bright lights, and a Southern feel. Every bar on the Riverwalk open and inviting to a gaggle of drunken soldiers stumbling from one building to the next. Walking tall, proud of who they were and what they did.

He'd never been to Texas before, but he knew what home looked like when he saw it.

Medic school took off running the next week. PT in the morning, eight hours spent in a classroom working on EMT certification, and nights spent studying. The weekends remained largely the same; Diego and Jay took others out as they damaged their liver and chased women and good times on the town.

In one such bar, the two of them watched all the pretty people dance while they sat at a table and drank. "I tell you, man," Jay said. "These kids have no discipline. They walk around like they've already been to war and back three times, and the NCOs don't let us do shit to fix it."

That much was true. The first sergeant had come out and said that corporal punishment had no place in his army. No more than ten years older than Jay, he paced back and forth in front of the formation.

"We make examples out of privates that can't conform here, we don't make them push, and we don't lay hands on them. You get me, prior service?"

Jay and the other veterans sounded off, but it left a bitter taste in his mouth even the beer couldn't wash away.

"I mean, what the fuck is the point of a military without discipline?"

Diego waved away Jay's concerns with an eye roll while he flagged down the waitress. "I swear, being infantry is like laying a fart so nasty you gag yourself. You guys go on and on about how hard the military has to be to work, but it doesn't. You build things up way worse in your head than they need to be on the ground, and all it does is fuck you up in the melon."

Jay seldom gave Diego shit about being a POG, but it was his turn to roll his eyes. "And what do you know about it? You cook the meanest eggs in all of Baghdad and you think that shit makes you tough? You don't look hard when you're suckin' on a juice box, scrub."

"Shit, bitch." Diego took a playful swing at Jay, much the way Nelson used to. "Hard don't mean a fuckin' thing. It's all about what you get out of it. If it gives you nothing but a headache and a bad time, what the hell good did it do anyone? It certainly didn't help Iraq, did it?"

A sore point if there ever was one. If anyone but another veteran had said it, Jay might have thrown his beer at the man instead of sipping it and peering over the bottle.

But Diego wasn't done. "Get what you can. I'm getting this shit to help me get into medical school when I get back to Georgia. Full stop." He polished off his beer and shook the empty bottle at the waitress. "And you're spending your time…what? Because getting mad at these kids for not knowing their head from their asses is stupid. You might as well get in a time machine and get mad at yourself for that. You were no different. I wasn't."

Jay's temper flashed red before a laugh erupted from him. He'd missed that blunt, easy honesty that soldiers wore so well. "Just that easy?"

"Just that easy, man."

They kept drinking, the subject turning from the privates to the war and back to all the beautiful women in the room. A few hours later, they cabbed back to Ft. Sam, sleeping off a Friday drunk before waking up to start in on the Saturday one. And why not? As Diego had said, such was the life of the soldier. The search for deeper meaning or purpose in any of it was only as far away as the next order.

Or so it seemed when he was drunk and with friends. But when he returned to his bunk at night, it wasn't so easy. Charlie Company has been predicated on the idea that taking responsibility for the world and the unit around you made everyone better. It hadn't just gotten them through fifteen months in Iraq; it had taught Jay to deal with everything that came after. If everyone had only been in it for themselves, things would have been different.

By Monday, the questions began. Wondering if there was a point to being here. If there had been a point to being anywhere in the army. It had forged him into the man he was, but was that it? Was he no different from Diego, only here for himself?

The privates disrespecting everyone and everything around him didn't help. It was like babysitting without the pay. Had he been as undisciplined as the little bastards in his class? The compulsive liars? The other specialists who thought they were equals because they were the same rank? Young men and women who barely had any clue what ranks were, let alone how to respect them.

The NCOs in charge, half of them without combat patches, watched and did nothing. They played grab-ass like privates and talked about college during the down time. How the army would help the new soldiers get ahead. Every word spoken was about what the military could do for them, not vice versa. No talk of wars. No talk of civic service. Only how they would be hailed as heroes, many of them having never lifted a finger to earn that title themselves.

More and more, he saw the cracks in the machine that must have been there the whole time. How it had gone unnoticed for years? None of the privates seemed to know why they were there any more than he did. Some spoke of patriotism, but he'd done the same as a new private only for Nelson and Duarte to laugh at him.

Everyone in the military professed to love their country, but few joined because of that.

The last straw came only weeks before graduation at a company formation.

First Sergeant paced back in forth in front of the assembled crowd. "We've chosen Private Jessup as soldier of the cycle. She'll receive a certificate, and she'll give a speech at your graduation. Everyone give her a hand."

Benjamin did as he was told, but his guts dropped. A mediocre student, Jessup worked half as hard as others and received half the results because of it. There was only one thing that set her apart from the bulk of the students, and First Sergeant confirmed it with his next few words.

"Her father, Sergeant Major Jessup, will be speaking at graduation as well."

He put out the schedule for the upcoming week and talked about the events going on around base, but Benjamin stopped listening. This wasn't his army. This wasn't any army. He and his had fought, blood and bullet, to serve their country. To make the world better or to punish those who'd hurt their home. And while they'd all had ulterior motives, the mission always came first. They placed what needed to be done, no matter the cost to self, before any private affair. Merit was earned by the individual, not their family.

So what the hell was this place? As they broke formation and returned to their quarters for the night, the question bounced around his skull. It spun faster and faster until it hit a velocity where it couldn't be seen, only felt. An uneasy headache setting in. A growing fear that it might all have been for nothing, that maybe he'd seen it wrong at the time, and *this* was the real world. The real army.

The privates wore the excitement for their new lives on their sleeves. Everything here must have looked like golden opportunity to them the same way it had to him when he was new. But he wasn't so new anymore. The more he listened to them speak, the more he heard the lies their recruiters had told them, the more he wondered why anyone bothered. What was it all for? Was it all just opportunism? Was there no greater meaning?

The classes wore on. Diego joined Jay drinking every weekend. They laughed and sang, dancing once they were drunk enough. They took pictures at the Alamo, and did mock shoot houses during work hours to practice casualty evaluation.

Blood covered everything. The air smelled of fake smoke pouring from the machine in the corner. Strobe lights flashed off all around. The high-tech dummies screamed bloody murder.

The NCO behind Benjamin wearing NODs yelled in his ear. "Move! Move! Move! Get in there and save who you can before your position is overrun."

Benjamin jumped to work. "You two!" He pointed at two privates. "Check the back rooms." He pointed at another two. "You two pull security on the door. The rest of you start triage."

Iraq had been hot, sticky, and smelly too, but not like this. Security had been different as well. There would be high sight to consider. Aerial reconnaissance. There would be a platoon, if not a company, extracting this many casualties, and the risks would be higher. Bombs under the floors waiting to kill the first man to step on them. A sniper half a mile away already running by the time the bullet pierced your skin and cracked your bones.

If there were no stakes, there was no value in the training. For him and Diego, months and months spent overseas had raised those stakes.

This? He stared down at the dummy as it moaned at him. Obviously fake, no matter how high-tech its squirting blood and realistic sounds might be. He looked around at the privates working around him. Terrified. Gripping fake plastic rifles and shaking as if their lives really depended on this, or maybe just their grades.

But he wasn't in the army anymore, just pretending. He wouldn't go back overseas. The medical reserves deployed in country to help units that *were* going to war get ready. They didn't fight battles or save lives. The most he would do was draw blood or run paperwork.

The NCO behind him kicked the back of his vest. "Get to it, Benjamin! You're killin' your patient with your lollygaggin'!"

He did. The sticky, red nectar that covered his hands smelled like food coloring. The stink of the fog machine didn't compare to the burning trash in Iraq in the least. When they finished, they stepped outside to hear the review of their performance. A cool breeze had set in on a chilly November day. Instant relief. No cloying, energy-draining heat all the days of his life. No stakes. No concerns as teenagers high-fived one another for doing nothing.

"Good job, privates. Very well done. You saved a lot of lives today."

Benjamin did his best not to roll his eyes.

That night he washed away the fake blood. It turned orange in the shower, circling the drain and vanishing with the brown dirt and flakes of mud. Real blood looked different on the ground and in a sink.

This was supposed to be a return to form. To something familiar. An escape from the trivialities and inane bullshit of civilian life to something that made sense. Instead, children ruled the roost as if it was fucking Neverland, and the NCOs were too busy trying to polish their own careers to care.

If that wasn't the truth, he didn't know what was. Everyone worked for themselves here. By the time school had been going for two months, the impatience he'd felt just before he got out the first time had replaced the thrill of being back. He whiled the days away studying, and spent nights drinking with the other veterans. He tried not to think about why he was here, why any of them were. That line of thought led him down a bad road. After all, if the Devil lived in the details, then he was in the big picture too. If the soldiers were only in it for themselves, what did that say about the leadership? The politicians that had sent them to war? The good he'd thought he'd done in Iraq, small as it might have been? Had he been fooling himself all along?

They rucked out to the field and stayed a week-long in the cold. More dummies. More privates playing soldier. More of that good army training that wasn't good after all.

A week later, tired, sore, and hungry, they marched back to the barracks. Orders were given, accountability taken, and within hours, they were back in the barracks, showering away a week of filth and talking about the plans they had for the weekend.

"You coming, Benjamin?" Diego threw a water bottle over the top of the shower and hit Jay in the head with it. Many of the people out in the field had treated it like a game, and he'd done the same in the real army. But this was different. These weren't soldiers, just children playing at it. They hadn't earned that yet. They hadn't earned anything.

"I think I'll stay in."

Diego laughed. "Your call, you grumpy bitch."

Diego probably had a point with the jab, but Jay let it slide. He opted to read a book instead of going out with the veterans who now freely mingled with the new blood.

Maybe he was just being an asshole. Maybe he looked at the military all wrong. Either way, the feelings wouldn't abate. By the time the graduation ceremony rolled around the next week, he was well past ready to be done with Fort Sam.

He sat next to Diego until they were each called in turn. They walked one behind the other up to the podium to shake the hand of the CO, the First Sergeant, and Sergeant Major Jessup.

"Well done, soldier, well done," the Sergeant Major said.

They filed back to their seats and listened to speech after speech about how these young soldiers would go out into the world and make it a better place, but Jay had his doubts.

A half hour later, he packed his car to drive back to Tennessee. Diego joined him as Jay closed the trunk.

"You out, homie?"

He pulled Diego in for a hug. "You know it, man."

Diego embraced him before holding him out at arm's length. "Maybe don't be such an infantryman now that you're a medic. You look like someone just shit on your breakfast table."

He felt like it too, but he kept the thought to himself as he put his sunglasses on and walked to the driver's side door. "Take it easy, brother. I'll see you on the other side."

"Love, peace, and chicken grease, soldier-medic."

It was the last time Jay ever saw him.

Chapter 11

Whenever it rained, the desert became an ocean. It poured in the cold months, and the sand became mud that caked on boots and tracked through tents, stealing the shower shoes off your feet.

And it was cold. So damn cold. He'd grown up in Michigan, but the drive from Mosul to Baghdad was the coldest he'd ever been. They'd pulled a mission on the way down and ended up sleeping in their trucks as a company. He'd shivered and shaken along with the other mortars. Waking up in the middle of the night unable to feel his fingers or toes.

"Turn the fucking heat on!"

Nelson threw a full magazine over the mortar and hit him in the helmet as a response.

He'd been banished back to his own squad, and the RTO torch had been passed—not out of punishment, just change. Positions often rotated for the sake of it. Nothing made the army's dick hard like unnecessary bullshit.

They reached the heart of the war, BIAP, the next day. Huge was an understatement. The Baghdad Airport had been converted into a series of FOBs that could have fit half a dozen Marezs and Dimaondbacks[34] into the space it filled. It took twenty minutes to drive to the gate and exit from their quarters.

After the drive down, endless procedure followed. Tents to set up. Paperwork to do. Supply depots to fix up. They sat on their hands for a week, doing nothing but busy work and wondering when the next mission would come along. Their nice CHUs in Mosul gave way to a tent that fit half the platoon. No longer just him and Sutter, now it was all the mortars except Miller, who stayed in the senior NCO tent with Garcia.

It took some getting used to.

"I want the top bunk," Simmons said. They'd been standing at Nelson's cot arguing about it for an hour. "Benjamin is a piece of shit anyway. Fuck that guy."

Slap! Benjamin struck him across the face without a word, stunning everyone in the tent into silence. "I'm sorry, I didn't hear that."

"Benjamin, if you do that again, I'll—"

Slap! "You'll what, private?"

Simmons squared up, but Nelson, who'd been ignoring them, jumped to his feet. "Pushups! Both of you dumbasses."

---

[34] The two FOBs in Mosul. C Company stayed on Marez.

The others snickered. Benjamin dropped and began to push, but Simmons wanted to argue. "Sergeant! He—"

"I said push, motherfucker!"

They pushed beside each other as Nelson made his decree.

"Benjamin is the higher rank, so he gets the bunk."

That was that. Truth was, tensions had been running high ever since the move. A lucky AK round through the cockpit had killed the helicopter pilot who had shadowed all their missions in Mosul. A fluke to be sure. Someone spraying and praying it would hit, only to have Allah answer their wishes. Two men down so far, and right before a stressful transition.

Being freed into the battlespace came as a relief. Infantry were carnivores, and like any carnivores, they needed to hunt. They needed prey to stalk and things to devour. It was a new place, and more dangerous than Mosul, but that only made it more exciting. Soon, they laughed and joked just like up north.

They still ran patrols, but there were more raids now. Going out into the depths of night, looking for one man. It didn't help that the descriptions were all the same. *BOLO*[35]*: Middle-Eastern man between 5'8 and 6'2, medium build. Dark skin, dark hair, dark eyes.* The platoon always joked that they were describing Duarte, who fit the bill perfectly.

Their success record spoke for itself. After every mission, they explored their surroundings. They did the Iraq tour. Come see

---

[35] 'Be On the LookOut'

Saddam's palaces with the golden toilet seats. See what had once been priceless works of art, now little better than rubble. See what it looks like to be part of a war where the support staff saw movies on big screen TVs while the people who fought lived in tents with a dozen other men.

The last part crawled under his skin more than the rest.

They drove their Stryker all around the man bases that composed BIAP. The main base, Victory, had a PX the size of the one back on Ft. Lewis[36]. More people milled about it than at most Walmarts back home. They walked around as if they weren't in a war zone, but all were armed to the teeth.

The juxtaposition struck him no matter how often he saw it. Two young men, armed to the teeth, carrying armfuls of McDonald's to a Humvee. A young woman doing duck lips and holding up a peace sign in front of the early warning towers, her friend taking her picture. The biggest military tent he'd ever seen surrounded by sand bags and covered in advertisements for cheap televisions and video games.

The mortars parked their Stryker in the dirt lot outside the PX as two people walked out carrying a big-screen TV.

"What the fuck?" Benjamin muttered under his breath.

"What?" Olson climbed out of the TC[37] hatch and joined him on the ground.

---

[36] Army base in Washington State where C Company deployed from.
[37] Track/Tank Commander

Benjamin shook his head. "Is this a war or not? Are people seriously buying TVs and playing Xbox while we get blown up outside?"

A woman sneered as she passed. POGs from front to back. All of their uniforms were spotless, while his was dirty and worn in comparison.

Sutter and Donner jumped out of the vehicle last, and Sutter patted him on the shoulder as he walked toward the entrance. "Capitalism at its finest, buddy."

He played Xbox and smoked between missions too. He ran to the phone center to call home every chance he had. Only the violence of his job separated him from them. He tried to remind himself that enjoying free time wasn't a privilege. But Taco Bell and Pizza Hut in a war zone made it feel less like a battle and more like the world's worst vacation. Finding deeper meaning in a conflict didn't come easily with mascots present.

Regardless, they did their job. They fought their battles and came home victorious—or at least unscathed—every night. Come Christmas day, instead of exchanging gifts, they took part in a battalion-level mission guarding some filthy section of road downtown.

Olson stood in the TC hatch while Benjamin manned the neighboring one with the 240. Donner sat in the driver's while Masters and Simmons waited in the back with Sutter for the combat that never seemed to arrive.

"This is fucking dumb," Benjamin said after hour four. He couldn't remember why they were here. Door-to-door knock and searches? Target acquisition? The most over-the-top presence mission ever? It didn't matter. "I've got to take a shit. Who's on deck down there to catch it? Open wide."

Sutter sighed loudly into the mic. "Benjamin, shut the fuck up."

They repositioned the vehicle every so often in case anyone scoped a weapon on them, but the view remained the same. A long street through the middle of downtown Baghdad. Shops, mostly closed, lined the road. Cars everywhere. More people than he had imagined were in all of Iraq. All gave the ghosts and their Strykers a wide berth. They stared but dared not approach.

Packs of children ran around, feral and angry. Teenagers plotting with their parents to kill as many as they could—not just Americans, but their neighbors. Their people. Anyone who wasn't exactly like them. It brought bile to the back of Benjamin's throat and anger over his heart that burned so badly it hurt.

A tall, lanky Arab boy no older than fifteen stared at him. His threadbare clothes hung loose on him, and his gaze could have burned a hole in the Stryker. Benjamin stared right back, though he should have been looking everywhere for signs of a threat. Did that young man hate them? Did he blame them for the totalitarian nightmare they'd brought onto this country? *Be in by sundown, or your life is forfeit. We can bust in the door to your house as we please.*

Benjamin sneered and glanced away, scanning the road around him. Broken and beaten buildings all around. Not quite rubble, but the quality control department for the Iraqi building inspectors must have been on permanent hiatus. The cars parked around him were all older than anyone in the Stryker. Everything showed signs of wear from years of a brutal regime followed by years of occupation under a foreign army.

"I hate this country," Olson said. "I wish we'd never come here."

"Miss your wife?" Donner asked.

"Yeah."

"I miss your wife too."

The whole Stryker cracked up laughing.

"But for real," said Olson. "What the hell are we doing here?"

A good question, though Benjamin couldn't say if he meant this mission or this country. "We're doing our jobs."

Sutter groaned into the headset. "Shut the fuck up, Benjamin."

Same song, different day. The company and platoon radios crackled off and on. Miller and Nelson checked on them constantly, not fully trusting a corporal to be in charge of a truck. Miller never trusted any of them, even after spending years in their company, or maybe because of it.

"What do you want for Christmas, Masters?" Simmons asked.

Masters had been sleeping, but he didn't miss a beat. "I want to go home."

A sentiment they could all agree with.

"I could shoot you in the leg. You'd probably go home pretty quick after that."

Benjamin couldn't see what was happening inside, but someone picked up a rifle. "Here you go. Make it a through and through, and don't fuck it up."

Simmons laughed. "Corporal Olson, can I shoot Masters?"

Olson seemed to consider it. "No. If you shoot the one black guy in our squad while I'm in charge, I'm never going to hear the end of it."

Everyone laughed. "Affirmative action screws us again," Simmons said.

The sound of a slap came from below. "Fuck you," Masters said.

"Why the hell does everyone keep slapping me?" Simmons yelled.

"Maybe it's because you're an obnoxious cunt." Benjamin spit off the side of the vehicle and stared at the young boy once more. He hadn't moved. Just kept watching the Americans with hard eyes that had no place in the face of such a young person.

It carried on like that all day, through pointless radio chatter and the even more pointless mission. Operations like this made it hard to accept there was a greater purpose to the conflict. The captain had once remarked that war was like going through a brick wall with a feather duster. It would take time, and they wouldn't notice the progress as they made it. But the more Benjamin thought about it, the more he realized the advice had been bad. You couldn't get through a goddamn wall with a feather duster.

The mission ended by sundown. Not long after, they were back on the FOB where they would bunk down for the night before returning to BIAP.

"Merry Christmas, fuckers," Nelson said as he hopped off his Stryker once they all parked. "Let's go get some chow."

It didn't feel like Christmas. Dirty men stood in a parking lot with buildings and T-barriers all around. They passed a tank hit by an EFP. He could see straight through it to the walls of the building on the far side. It had almost certainly killed whoever was inside.

Terrifying, but of little consequence. War was awful; they didn't need to see any more of it than they had to know that. Hunger and thoughts of home and holidays gone by pressed harder as they sat at a table in a chow hall surrounded by other dirty soldiers.

A band had come from some other base to play music for the troops out in the battle space. They played military songs, Christmas songs, and last they played Mario Brothers music. The POGs all around them cheered, but the Chargers looked less than thrilled.

"That's fucking obnoxious," Sutter said to nobody in particular as he shoveled more rice into his mouth.

Others nodded or grunted their agreement.

"Think they take requests?" Benjamin asked.

Boom!

The first mortar landed somewhere nearby, the reverberation too hard for it to be far away.

Boom!

The band panicked, several of them dropping their instruments. One dove for cover, and another screamed. Before the next round fell, all of them had abandoned the stage.

Boom!

Across the chow hall, the POGs jumped under tables as someone cheered. It had been one of the Charlie Company guys. Explosions detonated around them every day. It wasn't hitting the building, and they were fucking hungry.

Boom!

With every blast, a wave of grunts and hollers rose from the tables where they sat. Soon they grew into cheers, a chorus of hollers with every round dropped. By the time the barrage ended, only they continued eating as the others in the chow hall looked on in shock.

Chapter 12

"So you were like, overseas?" the young woman who sat next to him in Spanish class asked during their break. "I have a cousin who spent time over there. Do you know Jody Ziegler?"

He did his best not to roll his eyes. Half the time people found out he was in the army, they wanted to know if he knew their friends. As if it were a small social club and not an organization dedicated to killing the enemies of the United States. "Can't say that I do."

She kept talking while he finished his cigarette, but he ignored her. Nothing had felt right since he'd returned from medic school. Classes went well, and he was working in a hospital instead of wearing polyester pants to work, but something was off. That sense of belonging, gone for so long, ached inside.

Olson and Donner had ended up in the same unit after they'd decided to stay in. Both had returned overseas. Two of the three guys he'd grown up alongside in the army put their lives in danger while he listened to undergrads blather on about patriotism. It made him sick. The people who said, "Thank you for your service" or questioned him about Iraq as if they were fulfilling some civic duty were kidding themselves. Little platitudes like that allowed them to go on thinking they were doing something, anything, to show support for a war most understood only as much about as what they saw on TV.

He finished his smoke and flicked it away. "I better get back inside," he said, interrupting her mid-sentence.

Her mouth dropped open a bit, but she nodded. She didn't try speaking to him again.

He woke up the next day to the same routine. Many of the other Chargers said the same thing when he spoke to them. It felt aimless. Life after the army lacked the direction and purpose they'd once had. Everyone fawned over how difficult being at war must have been, but truth was, living on the outside afterward was much harder. Once, he hadn't needed to find his own direction.

Outside, everyone else was just as misplaced as he was. More so, maybe, because so many people his age had no concept of what they lacked. There'd been lost generations before. Is that what this was? Living in an age of miracles, an age of instant communication, space travel, lasers, and medical marvels, had everyone become jaded? Had the fantastic become so commonplace it was now the mundane? Was the outcome of comfort always apathy, or was it a world gone mad not because of civilization, but because it was still a jungle pretending to be otherwise?

Was all of his discontent just a product of that, or was everyone as confused as him and his friends?

Some had gotten out to make something of themselves. Firefighters, nurses, doctors, police, activists—a member of Charlie Company could be found in any time honored profession.

Next drill, he spoke to his NCOs about transferring to a unit that would deploy overseas.

Master Sergeant Wells closed her laptop and folded her hands on top of it. A no-bullshit woman, she'd served long enough to remember a separate army for women. "You're about to get promoted. You just got through medic school. Now you want us to send you away?"

He sat down in the chair across from her. It was very easy to be casual with these people. Hard to imagine he'd been with them nearly two years now. This place had always been a backburner. He didn't even recall half their names when drill was up. "I want to actually be in the army, or do something like it. This?" He waved around at the office. "You know this isn't the army, Master Sergeant. This is something else."

"You don't think what we do is important?" She looked down her glasses at him.

A baited trap if he'd ever seen it. "It is, but it isn't important to me. If I'm going to deploy, I want it to be outside the United States, not down at some base I just left in Texas. I want to be back where all my friends are, not surrounded by a bunch of people who don't want to be here."

She pursed her lips, but what could she say? Most of these people had joined the Reserves to avoid the risk of call-up in the IRR[38]. A quarter of them didn't show up on any given drill. They never accomplished anything. No training, no classes, no nothing. It was as pointless as anything he'd ever done wearing a uniform, and she damn well knew it.

"Have you considered going as a civilian?" she asked.

He didn't know what he'd been expecting, but it wasn't that. "You mean as a contractor?"

"I do. The colonel could release you from your remaining obligation to the reserves, and you could go overseas like that. Or I could get you back into active duty, if you prefer that route."

He most certainly didn't want the second one. "He can do that?"

"You've got what? A year left on your contract? Are you going to re-up after that?"

He shook his head.

"Well, there you go. You don't want to be here, and you don't have to be."

He hadn't expected to be making this decision, but he was eager to make it nonetheless. "Yes, I would like that."

---

[38] Inactive Ready Reserve

Just like that, the wheels of progress spun once more. The weekend ended in a daze, and by Monday, he was searching through contracting jobs. A few years ago, he would have called contractors war profiteers. They made obscene amounts of money on the government dime, and the higher up you went, the more likely it was because of who they knew.

Maybe he'd had them pegged wrong, or maybe he was a hypocrite. Something needed to change. He couldn't last like this. Every day he felt as though he were climbing up a wall. He'd done everything he could think of to make it stop. Change diet. Exercise more. Focus on something else. Go out and talk to friends.

None of it had helped, perhaps the last one least of all. It was hard even to talk to Allen, his closest friend and brother, about the things under his skin. For too long the infantry had taught them you don't air your dirty laundry. You don't rock the boat.

"I'm thinking of becoming a contractor," Jay said as they drank at Allen's house one night.

Allen stopped with his glass of rum and Coke halfway to his mouth. "What?"

"Yeah. I've been considering it for a while now. The reserves will let me go, and I need to get out of here for a bit."

He set his glass down. "What the hell brought this on?"

What to tell him? That the daily grind was too much? That he thought constantly about his friends still in, or the country they'd lived in for so long? That he wondered if there was more to life?

That the most important thing he'd ever done ended before he was twenty-two?

"I just need a change, man."

Allen stared at him for a moment that was hard to bear. "I get it."

An uncomfortable silence filled the space between them. "Yeah?"

"I do, yeah. I've been thinking about that stuff a lot, too. But what's a guy with kids and responsibilities gonna do?"

Jay shrugged. "Take a drink and move on?"

"Take a drink and move on. Cheers, buddy."

They clinked their glasses together, and that was the end of that.

It was easier to hear he wasn't alone in lacking purpose. Most people did in their twenties. Maybe veterans more than most, maybe not. His other buddies implied it over the phone, but to hear it in person lifted a weight from his shoulders, at least a little.

The first callback from a medical contractor overseas came only a few days before the next month's drill.

"Hello," he said as he picked up the phone on his way home from work.

"Is this Jay?"

"Speaking."

"We hear you're looking for a job."

Chapter 13

That Mosul had turned into the predicted dumpster fire didn't need to be said. It had hit news back in the United States, which meant people asked him about it every time he called home. Miller announced it after a platoon sergeant meeting one day.

It did nothing to help the platoon's sour mood.

"Great. Well, good to know we wasted four months of our lives and a few of our friends to clean the place up." Benjamin sat on his bunk, shaking his head. They'd called it before they left, but that didn't make it any easier.

"Shut the fuck up, Benjamin." Nelson slapped him in the back of the head.

Miller stared daggers at both of them. "The shit's over. Doesn't matter now. We're here. Keep your head in the game."

Miller finished his announcements, giving them orders to be in bed early so they would be awake for the mission they needed to run at two the following morning. That translated to, "Go fuck around for a few hours," which privates and specialists did with gusto.

"This is fuckin' stupid," Masters said as he, Sutter, and Benjamin walked the five hundred meters from their tents to the nearby enclosure where the McDonald's and Pizza Hut were.

Even in the middle of the night in an out-of-the-way corner of FOB Stryker, Humvees drove by, people walked everywhere, contractors worked endlessly. It was a far cry from Mosul and twice as stressful. But dust still coated them. The smell of burning garbage tinged the air. It was the same in many ways, a constant reminder of how far afield they'd found themselves.

"Yeah? Well let's just go home then, man." Sutter laughed at his own joke, but no one else did.

"I'm serious. Why even do this shit if it's just going to get fucked again?"

A salient question. Talk had already made rounds about the election coming up next year and questions about the war. Would it continue until Iraq was stable? Could we just pull out soon? Was it worth the cost in lives? Nobody wanted to commit to it anymore, and the loudest voices were the liberals trying to get the United States the hell out of the Middle East.

But anyone on the ground could have told them Iraq wasn't ready for that, though whether or not anyone on the ground cared was another matter.

"Fuck it, man," Sutter said. "Just go stuff your face with some pizza and you'll feel better."

Pizza might lift their spirits but not by much. Even that was alien. Indian men who spoke only enough English to discuss the finer points of crust options stared at them, as if any moment the guns they carried might go off by themselves.

Caring about any of this grew harder every day. The rumor mill ran wild with talk of an extension to the deployment. The most common guess was they'd be there for two full years instead of one, but no official word had come down. Their future rested at the discretion of some politician-general who sat his lazy ass on a couch somewhere in DC.

Benjamin held out hope. No extension. A year and done, no more. It might have been unrealistic, but he didn't care.

"I'm gonna eat the fuck out of my feelings," Masters said as they ordered.

The POGs sitting at a nearby bench sneered at them as Sutter and Benjamin laughed. There had to be a sense of humor to everything here. They weren't going to survive without it.

Simmons ran to them from the direction of the tent, his breathing labored, his rifle held in his hands so it didn't sway on its sling.

"Mission, guys. Sergeant Nelson wants us all back right now."

"What!" Masters yelled. "I just ordered my *got* damn pizza!"

Everyone in the T-barriers around them turned to watch the drama unfold. Some were obviously POGs; others were guys from their unit or infantrymen Benjamin recognized from elsewhere. He didn't give one fuck about what they thought. "Yeah, dude. I paid for it, I'm eating my pizza."

Simmons rolled his eyes. "It's a company mission. We gotta go."

Sutter shook his head and started walking back toward the tent, but neither Masters nor Benjamin moved.

"Nah." Masters crossed his arms. "Fuck that. I paid for this shit, I'm gonna eat it."

Benjamin pointed a thumb at him. "What he said."

Simmons ran back toward the tents, yelling over his shoulder, "Fine, I'll let Sergeant Nelson know."

Masters and Benjamin locked eyes without saying a word. The missions had ramped up over the past few weeks. In Mosul, there'd been evenings that involved nothing more than quiet QRF, or only short patrols. Not anymore. Even the patrols lasted all day here. Nothing came easy. Little by little, piece by piece, whatever humor they had left was slowly draining away.

"Fuck!" Masters kicked a spray of rocks against the T-barriers.

"Hey, private, watch your mouth!" someone yelled behind them.

Masters and Benjamin walked back toward the tent without saying a word.

Simmons hadn't told Nelson anything, and he was too busy to care. "Get down to the vehicles and get them ready," he said to Donner and Simmons. "We SP[39] as soon as the company is good to go. Don't be the last motherfuckers down there." He rushed past Benjamin and out of the tent.

Donner prepped his gear as fast as Benjamin had ever seen him do it. "What the fuck is going on?"

Duarte already wore full battle rattle. "Some MiTT team guys down south got killed. A bunch of Iraqis dressed in US uniforms infiltrated their compound."

"Damn." Masters walked to his bunk and started getting ready. "That sucks."

"Yeah it does. Hurry the fuck up, you two. We gotta go."

The whole company rushed around the T-barriers where their tents were hidden. In no time at all, they geared up and ran to the motor pool. As fast as they were, most of the company had already arrived.

They called it theatre-wide QRF. Fast, armored, and lethal, Strykers were the vehicles of choice. The Patriot Battalion had been tagged for it. Only the best could manage.

"Fuck, man," Benjamin said to Duarte as they ran side by side to their truck. "This is crazy."

"It's a big fuckin' deal, man. A lot of guys got killed."

---

[39] Start Point

They'd prepped the truck so many times for so many missions that it came as naturally as breathing. Soon, Benjamin and Sutter sat in the familiar darkness of the mortar variant as Donner drove. Nelson and Olson manned the hatches.

"You guys good to go back there?" Nelson asked.

"Yeah, we're good, Sergeant," Benjamin said.

"All right, we'll be… What the fuck is this?"

The vehicle hummed and the engine roared, but nobody spoke. The suspense finally killed him. "What's happening?"

Olson sighed. "Pryor just jumped out of his vehicle, and he's walking back to the tent."

Sergeant Pryor was the leader of the company snipers. Leader was a bit of a misnomer. There were only three of them and they were all the same rank. Still, he called the shots.

"Seriously?" Sutter asked.

The platoon radio came alive. "He just fucking quit. I heard him from here," Garcia said.

They all joked about it at different points. Put your rifle on the ramp, head down to Combat Stress, and get that golden ticket home. It might ruin your career, but fuck it, right? At least you got the hell out of this circus.

Sutter laughed. "What a pussy."

Miller chirped in over the radio. "Enough."

By now, the company radio sang with chatter too. Captain Clinger cut it all off. "We'll deal with him when we get back. Charger Base, make sure he stays locked down with supply."

They left shortly after. Company formation on a drive down south. Keep your intervals. Keep eyes out. Pay attention to the radio and everyone stay awake. Benjamin watched the DVE[40] on the front of the truck through the FBCB2 monitor. The camera used to drive the Stryker at night painted the whole world in black and white thermal colors. They left BIAP and started on roads he'd never seen before. Within an hour, he was hopelessly lost. Within two, they could have been on the moon for all he knew.

"That was fuckin' crazy." Sutter covered his mic so the others wouldn't hear. "He just fucking deuced out."

Benjamin shrugged. "Guess Baldomero is in charge now."

"That guy was a piece of shit anyway."

Dismissal came easy once someone was on the outside. Like a pride of lions, they killed the weak and ate them. "Yup."

They reached their destination hours later, and the ramp finally dropped. Benjamin and Sutter had been asleep but woke up when Nelson announced they'd arrived. Outside, a floodlight revealed old stone walls slightly taller than he was. The same clinging, sandy dirt swirled through the cold January air here as in Baghdad. Old, brown stone comprised the Iraqi base around them, much of it ancient looking. Nothing was visible beyond the walls.

---

[40] Driver's Vision Enhancer: A thermal, night-vision camera in the front center of a Stryker.

Iraqi men stood around talking to the CO and an officer Benjamin didn't know. Maybe one of the MiTT team guys who hadn't been killed.

"Is this the place that got attacked?" Benjamin asked Nelson as they both stepped out of the Stryker.

"No. We're getting these guys out because it isn't safe here anymore. Iraqis drove right the fuck in, and fuckers like these opened the gates in the place that got hit." Nelson nodded at the Iraqis around them.

"So why aren't we shooting them?"

Nelson glared at him. "Shut up and give me a cigarette, stupid."

He gave one to Nelson and smoked one himself, both gazing at their new surroundings. Watching for points of entry, on the lookout for attack. Such was life. That he'd left without his shaving kit that had his toothbrush in it was the more pressing concern.

He told Nelson.

"You better find a razor, dumbass. First Sergeant won't give a fuck that you forgot yours."

Only in the army could five soldiers be dead, could they be out in the wilds of Iraq with nobody but one another to look after them, and have the leadership more concerned with shaving than anything else. "Roger."

The MiTT team they came to extract shot the shit with the company. Benjamin stood well back and let the leadership chat.

When E-6s and E-7s got together with officers to talk, the best place a specialist could be was far away.

One of the MiTT team members, one of his old drill sergeants, glanced his way. Benjamin did a double take, but there was no mistaking the man. He looked like a linebacker, and his chiseled features belonged on a movie poster, not in a soldier's uniform. But he was the same man who'd once bitched to a gaggle of new privates that he'd rather be sniping Iraqis than teaching a bunch of morons.

"Holy shit." Benjamin poked Simmons in the ribs. "That's one of my drill sergeants."

"Great." Simmons took a drag from his smoke. "Why don't you go blow him?"

He passed on that offer, but once the crowd of leaders had dispersed, he approached.

"Sergeant." He walked toward the man but didn't get his attention. "Drill Sergeant Amden."

This time the man stopped and turned around. A smile crept across his face as he recognized Benjamin. "Private."

Even specialists had been called private in basic training. Benjamin smiled at the inside joke. "Never thought I'd see someone else from Echo Company on the other side of the planet, Sergeant." He shook hands with the man. "Jared is somewhere around here too. Remember him?"

"Yeah, I remember him. Goofy guy from Mississippi. Mouth always open like he's gonna catch flies, right?"

Benjamin laughed. "Yeah, that's him."

"Kennedy was in your class too, wasn't he? I saw him at a chow hall a month or so back on FOB Kalsu."

Kennedy had been Benjamin's platoon leader in basic, a college specialist bred and raised to be military. The kind of solider who made E5 as soon as he was able and went to Ranger School as a matter of course.

"Hell yeah! How is he?"

"He's good. E5, Ranger School. He's working with his brigade commander."

"What about you, Sergeant? How are you holding up with all this shit going down?"

He shrugged. The stoic infantry response. "What are you going to do, right?" They stood in silence for a moment. "I need to get back to work packing. Take it easy, Benjamin. Good to see you."

"Good to see you too, Sergeant. Stay safe."

He walked back to Donner, who'd been sitting on the ramp watching the interaction. He sat next to his friend and lit a cigarette.

"It's weird, isn't it?" Donner stared into the canopy of the night.

"What?"

"Running into people you knew before Iraq. I saw one of my buddies from high school on BIAP. He looked older."

Benjamin took a drag from his smoke. "I think we all do."

They spent three days at that outpost. Packing up the MiTT team and guarding the sector was a full time job, but one that never involved leaving. What had started out as a tense mission into the unknown turned into the most boring three days in Iraq. He'd left with almost no smokes, no toothbrush, and no razor. By the middle of the first day, he was using a brush from his rifle cleaning kit on his teeth, and a disposable razor from Howard on his face. By the dawn of the third day, his face was raw and the taste of CLP[41] wouldn't leave his mouth.

Sutter and the rest of the squad laughed and pointed while he brushed with the green and black toothbrush never been meant for teeth.

"Whatever, man." He spit a mouthful of water that tasted of cordite. "At least I don't have nasty chompers."

The lieutenants and the platoon sergeants slept next to everyone else in a small building in the corner of the compound. A guard always stayed on the truck. The whole company might be here, and it might be a "secure" outpost, but they were always on watch.

Benjamin spent the day staring through the holes in the wall around the base, looking out over the landscape. This far south, it actually looked like a desert, all sand and barren landscape. Nothing to see. No trees. No water. No enemy. But that didn't mean they weren't out there. When he wasn't staring, he was fighting over the two English-language books found on the outpost. A romance paperback and a western novel older than anyone in the company by a generation.

---

[41] Cleaner Lubricant Preservative

The shop on the outpost had only Pines, the awful local cigarettes. It sold candy bars with pictures of acorns on them. A nice change from MREs, even if the flavor was questionable. They ran the place dry on a regional energy drink called Wild Tiger. Rosenberg, Howard, and the rest of third platoon went nuts for the stuff.

"Made with real tiger balls."

"Tiger flavored."

"Makes me want to eat a fucking gazelle."

But eventually, the jokes ran out, and only the quiet remained. Just being out there, in the middle of the desert, on the other side of the planet from everything you'd ever known. Introspection came easy at times like those.

To avoid it, Benjamin struck up conversation with one of the terps one evening. A light sandstorm had moved in. Not the sky-blocking monsters that turned day into night, just a small one; a mist conspiring to blot out the little sunlight left in the day. It cast the glow of the stars in a brown haze, as if time had stopped here. As if this place, and the people in it, had entered into some sort of purgatory.

The terp set himself apart from everyone else outside the Strykers. He watched as the soldiers spoke with one another, or sat quietly in the cold. Benjamin had never really talked with the man outside of passing greetings, but he knew him from around. When he sat next to the terp, the man didn't as much as look over.

"Hey." Benjamin pulled out a smoke and lit it. "What's going on, man?"

"Hello," the terp replied in perfect English. Benjamin could barely understand some of them. He would have sworn others grew up with English as their first language. For all he knew, many of them had.

"Sucks getting stuck out here, huh?"

The man shrugged. "Stuck here. Stuck in Baghdad. Always stuck somewhere."

That was the truth. Silence filled the space, long and uncomfortable, before Benjamin groped for something else to say. "What do you do when you aren't a terp for the US?"

The terp flicked his cigarette away. "I'm a medical doctor."

Benjamin did a double take. For the first time, he noticed the wisps of silver in the middle-aged man's black hair. The crow's feet under his eyes. He looked older than Benjamin would have guessed. "No shit?"

"No shit."

"What got you working with us?"

The terp clenched his jaw, as if thinking about something he didn't want to, or perhaps biting his tongue. "My hospital was shut down when the invasion started. When it opened back up, terrorists came and told us that if we helped the wrong people they would kill our families."

The door-kicking riflemen spent more face time with the yokels, but he'd spoken with them on a handful of occasions. They all told a sad story. "I'm sorry."

"I work with you so I can afford the fake passports to get my family out of our country."

Benjamin had no idea fake Iraqi passports existed, but it made sense. "Wow, man."

The silence that followed felt hostile. This man didn't want some kid coming up and asking him questions. But it wasn't Benjamin's fault this place had gone to shit. He hadn't been the one who talked about weapons of mass destruction on TV, or put a madman in charge before that.

He said the first thing that came to mind. "So, do you like working with us?" His face burned with embarrassment at the question.

The terp leaned out and took a good look at Benjamin, maybe seeing him for the first time. His features softened a bit, but he was still hard, still a man who had watched his country ruined. "No. Saddam was a monster, but at least I didn't have to worry that my daughter would be killed on her way to school when he was in charge. You Americans, you ruined this country."

He stood up and walked away, and Benjamin watching him in stunned silence. What could he do? Grow angry and argue? Fix the man with puppy dog eyes and implore understanding?

Anger did follow, but he kept it to himself as he smoked. The terp might not want them there, but the Kurds in Mosul had been glad to see them. Others had thanked the Americans for coming here and trying to help.

But that thought bled too easily into another. Their mission wasn't to help people. They did what they could, but the US hadn't mobilized just to depose a dictator. If that were the case, they'd be flying all over the world every other month. This had been something else, but he didn't know what.

The thought made him squirm, and he slept poorly that night. After all, if the Iraqis didn't want Americans in their country, if they didn't want the "freedom" American claimed to give, why the hell were they here?

The snoring of his brothers did nothing to answer his questions, and the howling of the wind only deepened the mystery.

Chapter 14

His nerves rattled even before the plane touched down. He resisted the overwhelming urge to order a drink to calm them. That wouldn't make for a good first impression. He'd been informed during the interviews that Kuwait was a dry country. He let go of the desire for one more good, hard round of drinks.

All around him, people spoke in a litany of languages: French, German, Spanish, Dutch, Arabic. It increased when he stepped off the plane and into the crowded airport. He'd seen it before. When they'd come back from Iraq, their plane had stopped in Amsterdam. He'd tasted his first legal beer in this airport.

Just nerves dragging his mind that far back. He took a deep breath and tried to calm his rambling mind, but it did no good. He dodged through the crowd, looking for his connecting flight. Adrift in a sea of faces, nobody paid attention to him. Something about that calmed him. Nobody cared who he was here. Nobody noticed him or his problems. He was as free in a crowd as anyone ever was.

A chauffeur stepped in front of him as he passed a row of elevators. "Excuse me, sir? Are you Mr. Smith?"

Jay shook his head and kept walking. Being addressed brought a certain sense of grounding to the unreal scenario. He walked calmer, more confident, to the next gate. By the time he boarded a plane to Kuwait, his nervousness had ceased.

The job had become a whirlwind once the ball was rolling. After the first offer had come in, a dozen others followed. He took the first shooter position that had opened up. PSD[42] in Iraq. They'd sent him to Louisiana to qualify and train for the position, a dozen classes over a long weekend. Everything had been set up and ready to go, bags packed by the door.

But the call he finally received wasn't the one he'd expected.

"How old are you?" the woman on the phone asked.

"Uh. Twenty-four. I thought we covered that."

"Hold one moment."

One moment turned into twenty while he paced back and forth in his living room. He'd been in the army long enough to know when bad news was coming.

He wasn't disappointed.

"I'm sorry. This position is considered a leadership role. You need to be at least age twenty-five to take it. When is your birthday?"

It had been three weeks before, and he told her as much.

---

[42] Protective Security Detail

"We cannot give you this job. I'm sorry."

He'd kept a cool head through the entire process, rushing it along as much as he could. Even when people had been as rude and dismissive as they had been in the army, he swallowed his pride and showed his manners. This, however, broke the camel's back.

"Are you fucking kidding me? I quit my job because you told me my plane ticket would be coming within the next week. You're going to tell me five fucking days before it's time to go that you don't have any work for me?"

The woman stammered for a moment. "Hold, please."

The line came back with a project manager. "Henry Thompson speaking."

Jay explained the whole situation to the man.

"That does sound like an issue. You know, we have some base security jobs in Israel that would be good for you if you're interested."

They'd been trying to pawn the Israel job on anyone who would take it the entire time he'd been in Louisiana. Nobody wanted work that paid half as much with three times the hours.

"No, you know what? Keep your fucking jobs." He hung up. A mad dash to find any contracting job that would take him followed. After two days of frantic searching, he contacted the ambulance company that ran emergency calls on military bases in Kuwait. PSD jobs paid better, but beggars who cussed out contract managers over the phone certainly couldn't be choosers, especially when they'd already quit their job.

In Dearborn, the Detroit suburb where he'd gone to high school, half the population was Arabic. Every morning the call to prayer rang out from the mosque near his house. Half his graduating class had been Muslim. The Middle East wasn't the great mystery to him that it was to many of the people he'd gone to Iraq with. They'd seen it as something other, something alien. It had sometimes been that, especially when they were stressed, but often just different. Not bad, just not what they were used to.

The plane looked like it had taken off from East Dearborn, where most of the Middle-Easterners in town lived. Half the plane's occupants wore headdresses of varying colors. Most of the men dressed in long, white robes. Burqas covered many of the women, some of them adding veils. When the Indian flight attendant asked if he'd like a drink, she asked him in Arabic before English.

"Excuse me?"

She smiled and laughed. "I'm sorry. Has anyone ever told you that you look Arabic?"

"Does that mean you think I'm handsome?" he asked with a smile. The words slipped out before he could contain them. And why not? Go big or go home.

She smiled again. It was a very charming smile. "Water, then?"

His nerves jangled. The greens of the world below slowly gave way to the browns of the region that had tried to kill him last time he was here. Kuwait was a small country. There were bigger states in the US. He wouldn't be far from the place where his friends had died, where he'd almost joined them dozens of times.

It's safe. He had to keep reminding himself, as if the repetition would make it true. *It's safe.*

He'd done his homework before he took the job. Almost nothing happened in Kuwait. They were not only staunch allies of the US but a laidback country by Middle-Eastern standards. Still, imagination could run wild when you visited a country so close to one mired in war.

Below them, the vast expanse of the desert gave way to the city. Kuwait differed from Iraq. The farther north you went, the rockier the terrain became. This far south, the land looked like the movies. Dunes of sand. Nothing but a sea of yellow sand intersected by the occasional long, lonely road. The only people who populated the world in those desolate spots were the US military and the Bedouins.

The cities were just as spectacular. Buildings as tall as any you'd see in New York or LA. Countless cars and millions of people going about their lives in one of the most inhospitable places on the planet.

He jolted as the plane landed. The pilot spoke over the intercom, first in Arabic, then in English. "Welcome to Kuwait."

His mouth went as dry as the sun-bleached bones among the dunes, but his water was gone. People stood up, getting ready for the doors to open so they could greet their loved ones. He stayed seated until everyone else left, watching them pass by one at a time.

He stepped off the plane and into the airport. The rush of people wasn't as dominantly Arabic as he thought it would be. He'd read somewhere that only two percent of the population of Kuwait was native. It involved the way the laws applied to natives as opposed to visitors, but it had still been hard to imagine. Seeing was believing. White, brown, black. Asian, European, Arabic, and African—as diverse as anywhere you'd see in the states.

Getting his passport stamped took no time at all, and in another ten minutes, he was downstairs at the currency exchange counter, trading a few hundred dollars in US currency for what looked like Monopoly money.

"Jay?"

He turned to see a tall, handsome man in his late twenties or early thirties.

"You're Jay, right?"

Jay stuck out a hand. "You must be Jack. How'd you know?"

Jack shook his hand and flashed a smile. "It gets pretty easy to pick out the new people."

Jay thought he had a good grip on his anxiety. He wondered if anyone else had noticed. "Good to know. So you want to get me to this office so I can sign some paperwork and get to sleep? I'm beat."

Jack waved the idea away. "We can do all of that tomorrow. How about I just take you to the place where you'll be sleeping for now?"

Jay followed him out to an SUV parked by the curb outside. People all around spoke in a dozen languages. Smells both familiar and foreign assailed him, and even in the middle of the night, the heat made him sweat.

The man behind the wheel of the SUV looked familiar. Judging by the way he stared at Jay, the feeling was mutual.

The man's eyes lit up just as Jay placed him. "Holy shit," Jay said, "you were in 2-3, right?"

The man laughed. "Yeah! I was in HHC[43]. Weren't you in Charger Company?"

Jay jumped into the front seat and shook the man's hand. "Sure was. Jay Benjamin."

"Chris Stevens."

Jack entered the back seat. "You two were in the military together?"

"Apparently." Stevens started the vehicle and drove off into the tangle of cars that composed the downtown traffic. Even this late at night, the bustle was considerable. "You were on the '06-'07 deployment, right?"

"The whole long ass time, yeah. I remember you now. I would always hang out with Lars and Pac. You worked with the PA, right?"

[43] Headquarters and Headquarters Company

"Yeah, man."

They exchanges stories and pleasantries, but Jay found it hard to focus, even with the unlikely coincidence that had just found him. Outside, the world darkened as they drove onto a highway and away from the city. Out here, in the blackness, it looked just like Iraq. Filthy brown buildings dotted the side of the highway. The farther from the airport they drove, the sparser the roads became. Kuwait wasn't as dirty as Iraq, but it wasn't clean either. And dust, so much dust everywhere, even inside the car.

He hunkered down in his seat, his heart hammering in his chest. This was no Stryker, just a black SUV. If something exploded on the side of the road now, there would be no protection. Other Americans had lived here for years, no terrorists ever attacked Kuwait, but his fears wouldn't relent. He couldn't unsee the similarities between this and the war-torn country he'd once fought in. They looked so much alike that for a moment, he forgot where he was.

Stevens noticed. "Weird, isn't it?"

"Fucking bizarre." They passed palm trees swaying in the wind and the bongo trucks so common in the Middle East.

"Took me a while to get used to. It's better in the day. More people, more going on. At night, it looks just like Iraq."

Jay finally tore his gaze from the window and looked over at Stevens, who smiled in return. "So you haven't been hit by any IEDs, huh?"

Both he and Jack laughed. "Not yet." He rapped his knuckles on the fake wood of the dash.

Hearing that from another veteran, from one of his brothers, eased his mind. His heart still beat faster than normal, and he sat low in his seat, but it didn't seem as menacing. Before long, they left the highway and turned into the tangled buildings between the crowded the strip. Bright lights, lots of people. More activity than in all of Chattanooga, even at night. The energy put him further at ease.

"This is nothing like I imagined."

"Yup. You'll live out here in the city. Your base is going to be Camp Virginia. It's a small one way out in the middle of the desert. Quiet but pretty nice. You'll meet your bosses in the morning."

They pulled up to a huge apartment building and parked. Most of the fear had bled away. What was left was...excitement?

"Welcome home." Jack waved an arm at the building. "We'll give you a temporary room for now, so don't get too settled in. Within a week, you'll be working and in your permanent apartment."

He grabbed the two bags holding all his possessions worth taking and followed them into the building. Even the architecture reminded him of Iraq. The oddly square proportions of the buildings. The way some steps were bigger than others. The floral tile placed into the wall. So alien yet so familiar.

On the fourth floor, Jack opened a door for him and showed him into his temporary apartment. Big, with a great view of the street below and the rooftops across the street. He could hear the energy of the city out there, and it calmed him.

"I'll come get you in the morning, I know you're beat. There's another guy living here, but he's at his two days on shift right now. You won't see him for a while."

"Cool." Jay couldn't stop staring at everything. That jetlag had vanished, as had the fear. Giddiness rose inside him, threatening to bubble up into a laugh.

Stevens held out a hand, and Jay shook it. "If you're hungry, there's a great Indian place just downstairs. There should be a bunch of water in the fridge. Stay hydrated."

"I will, thanks."

"Take it easy, man," Jack said. "Welcome to Kuwait."

They left, closing the door behind them.

For a long moment, he stood there, listening to the city. Smelling the dirt in the air and the gulf only half a mile away. Even through the air conditioning, he could feel the heat of the stone.

Alive. That was what he was feeling. How long had it been since he'd cut loose from every safe harbor and ventured out into the unknown? Every person felt that need sometimes. Was that what had been missing from his life for so long?

He carried his bags into his room and laid them at the foot of the bed before lying down and staring at the textured ceiling above him.

The streetlight outside bled through the red curtains just enough to paint the room a dark crimson. The last call to prayer of the day began, the *adhan* sung across the city by the *muezzin*. A dozen of them spread across the many mosques, a hundred, all combining to create a ghostly layer of song. It called the faithful to drop to their knees and thank Allah for the wonders he had created.

Jay wasn't one of the faithful, but the song tugged something sweet and familiar nonetheless. First from home, then from war. He laughed at everything and nothing at all. Silently chuckling at first, then bellowing at the ceiling.

Why not? For the first time in a long time, he was in his element.

For the first time in a long time, he was home.

Chapter 15

"Shots fired, right side," Nelson said over the headset. "Stay low in the hatch, Olson."

They took near constant small-arms fire. Just a man peeking around the edge of a building with an AK and spraying a few rounds, or a sniper a quarter mile away in a window. Nothing they could pinpoint and destroy.

Six months ago he would have called it a nuisance. Now men had died, and what had once been exciting and dangerous was just dangerous. Fear didn't make them cautious, only aware of what was at stake. The life of a brother wasn't easily replaced.

"Can we just light that building up?" one of the first platoon NCOs asked over the company radio.

"Keep this fucking line clear," the first sergeant replied. "It ain't a gosh darn toy."

After three days on that miserable outpost with nothing to do but sit on their asses, they were finally going back to a FOB. Kalsu, some little base he'd never heard of twenty miles south of Baghdad. The news had been welcome, though it was hard to muster the give-a-damn to be excited. Nobody knew when they would be going back to BIAP.

The south became more unstable by the day, the attacks more frequent. Higher-ups on the theatre-wide level talked about troop surges, bringing in thousands upon thousands more soldiers to fight off the tide of enemies that were flooding into Iraq. Or the ones who had been born here. Or...who knew? Everyone wanted to kill invaders, and it became impossible to keep track after a while.

What the terp had said wouldn't stop going through Benjamin's head. They weren't welcome here. It didn't feel like a peaceful occupation, and it wasn't. This was a war. Iraqis didn't want them here. People back home didn't want them here. The UN didn't want them here.

But what was a soldier to do? They'd been raised by rhetoric, by Uncle Sam's siren song. Look what your fathers' fathers did! Look what they accomplished with bloody hands and gnashing teeth. You too can rule from the top of the mountain if you have the courage to claim it. Plant your flag on its highest point and laugh at the trail of blood that got you there. Until then, you're just a maggot. A speck. A filthy millennial who sucked the system dry.

The world turned them into faceless monsters and thanked them for it. They called it service. But it was always in comparison to the greatest generation. The one with the highest body count. The one who earned the title with the flash of a bomb, and saved the most lives with the business end of a bullet.

But times had changed, and the reasons for this war were muddier, even for those fighting it. Would the United States welcome them back and call them heroes? Would history remember this kindly, or would they be condemned by the generations that came afterward? He couldn't be the only one thinking about that. Plenty of people in the company waxed philosophical. Staff Sergeant Griffin of third platoon never missed an opportunity to discuss the nature of the world, war, or humanity. Others certainly wondered what the fuck they were doing here.

"You daydreaming about Biffany?" Sutter kicked him in the knee pad.

"Oh, Biffany!" Donner laughed over the headset.

Nelson never missed an opportunity to join in. "You're going to make a beautiful bride, Benjamin."

Benjamin rolled his eyes. "I'm not a traditionalist. We don't conform to gender roles in weddings."

Sutter shook his head. "Faggot."

"Just be careful," Nelson said. "I heard she has a clit like a dick. That shit will tear you up if you don't use lube."

They all broke down laughing, and in a moment his internal tension vanished. So easy to forget the golden rule: keep your eye on the prize. A year before they'd left, they'd done the EIB[44] test. Miller had given them that advice then, and it still applied.

"Don't think about the next station. Don't tell me how you did when you get out of it. If you have to do it again, say, 'One more time.' If you make it, say, 'Next station.' That's all I want to hear from any of you."

Every private in the squad passed because of their hard work and his advice, even when half the company failed. The guy might be an asshole, but he knew his job now as much as he had then.

"If I see one of you assholes counting down days on a calendar, I'm going to slap you."

Benjamin smiled while they drove through the darkness. Just more ghosts in a country full of them. Nothing but tires over broken roads and the not-so-quiet-hum of diesel engines to mark their passing.

They pulled through the gate of Kalsu an hour later. He could see nothing in the blind tomb that was the back of the Stryker.

Before long, they parked and the ramp dropped. Here, the air didn't smell quite as bad as it did out on the road or around Baghdad, but it was still dusty and kicking the burning garbage funk. In the darkness, it looked considerably more cramped than BIAP. Where there weren't more T-barriers there were giant cardboard containers filled with dirt playing the same role. The amount of blast scars from where mortars must have landed alarmed even his jaded senses.

It looked a lot more like what he imagined a war might.

---

[44] Expert Infantryman Badge

Apparently, Nelson agreed. "This place is a shithole."

Benjamin lit a cigarette and handed Nelson the pack. "Your mouth is a shithole."

Nelson took the smokes while trying to hit Benjamin in the head with his helmet. Before long, the whole company had parked and everyone climbed out of their vehicles. Tiny, scarred-up shithole or no, it was better than the outpost where they'd been living.

"First place I'm going is an internet café. I haven't talked to Tiffany in a week and a half." He'd put it off before they left due to the volume of missions. After being out of the wire for sixteen hours straight, he seldom wanted to trek to the phones. That it would bite him in the ass had been a matter of when, not if.

"No, you need to go wash your dirty balls."

"Me and Corporal Olson can go find the phones and PX and report back to you guys while you get all the bunk shit sorted out. Everyone is going to need to know where it's at anyway."

Nelson considered it for a moment. "Fine. But hurry the fuck up. I want everyone nearby if they start putting shit out."

They didn't wait to be told twice. Olson and Benjamin trotted off as soon as they finished the immediate work. Howard followed them with Lunquist's permission.

"You know the bullshit details are coming down." Howard adjusted his 249[45] and put on his worn, dirty soft cap. "I'm getting out of here before then."

---

[45] M249: Belt fed 5.56 machine gun.

"Sham shield[46]! Activate!" Benjamin rubbed the rank in the middle of his uniform.

It didn't take long to find the phone center. The rest of the FOB bore out the first impression; the place was a mess. More T-barriers than on FOB Stryker back at BIAP. More craters.

As they walked, the PA system came alive. "Incoming, incoming, incoming."

The first round struck somewhere distant enough not to concern him, but close enough that he eyed the nearest concrete mortar bunker.

"It's going to fucking suck here," Howard said.

The rounds rained down, ruining someone's day somewhere Benjamin didn't care about. Others did, though. POGs hid in the barriers, waiting out the end of the shelling. Benjamin, Olson, and Howard didn't as much as slow down.

The PA spoke once more. "All clear. All clear."

POGs as scared as Benjamin had felt in those first few days in Iraq peeked their heads out. It hadn't been so long ago, but it felt like a lifetime.

"Don't be afraid," Howard said as they passed a trio of privates. "First to go, last to know."

Olson laughed, and Benjamin smiled as he shook his head. The privates however, looked less than amused.

---

[46] The rank of specialist is denoted by a shield with an eagle on it.

They didn't wait long for the computers, and soon Benjamin had pulled up AOL Instant Messenger. By some miracle, Tiffany was online.

She messaged him before he messaged her.

"Oh my god, where were you? I've been waiting forever to hear from you."

He smiled as he imagined her saying it aloud. "Been busy. Sorry. We're on a different base now, and we were stuck in the middle of nowhere forever." He caught her up on what they'd been doing, but it wasn't long before he ran out of things to say. Soon only small talk remained, and even that ran out. Within fifteen minutes, he repeated the same platitudes he did every time they spoke.

"I love you. You're so beautiful. I'm so glad I found you." The saccharin sweetness always earned a gag from his teammates.

But eventually he began to worry, even if he kept it to himself. A relationship required some kind of affection, and that was something he couldn't give from the other side of the world. And so he sat, trying desperately to prevent the conversation from lapsing into nothing. Luckily, Olson saved him from that.

"Come on, man. Time to go." He tapped Howard on the shoulder as well.

"I love you, Tiffany. Don't ever forget that."

"I love you too! Be safe!"

They hadn't known one another long before he'd deployed, only a few weeks. Still, having someone who missed you when you were sweating and slaving on the far side of the planet was nice. It was good to imagine that someone would cry if you were gone.

It brought hurt to any night he spoke to her, but a good hurt. He needed the stability that came with having someone on the outside. They all needed it. The stresses of their jobs mounted with every mission. Day after day, week after week, and rumor after rumor. Everyone knew what nobody was saying; it was only a matter of time before someone else died. Before the next horrible thing happened.

"Benjamin! Come on!" Olson waved him toward the door. Soon they were back into the not-so-fresh night air, walking toward their temporary home.

"How's Angie?" Benjamin asked to fill the silence.

"Good, man. I miss her." Olson adjusted his soft cap as he always did when he was nervous. "Baby's good too. She's talking a lot now."

In the big groups, it was easy to be hateful, pretend or no. Someone would make a comment about his wife not missing him, or that he needed to nut the fuck up. It almost never happened in smaller gatherings. With just a handful of guys, it was okay to show that soft side for a moment.

"Not too much longer now, buddy. Just keep chanting 'No extension, no extension.'"

Olson smiled. "I am, man. I am."

"You boys might want to get ready for it," Howard said. "It'll make it sting less when it happens."

By the time they returned to the tents, everyone was already inside. Howard waved as he jogged toward the third platoon area.

"Looks about the same as Baghdad." Olson stepped inside.

It wasn't. In Baghdad, they'd had time to settle in. Bunk beds instead of cots, blankets, TVs, and a myriad of other comforts spread around their living quarters. It felt less like a tent and more like a home of sorts.

This was just a tent. Cots, a dirty floor, and nothing but the bags they took with them when they left BIAP. It was Spartan. The kind of living he thought he'd be doing when he joined the army.

Back then, he would have expected it; now it bummed him out. "This is where we live?"

"Shut the fuck up, Benjamin." Nelson slapped the back of his head as he walked by, shower bag in hand. "If you don't like it, you can sleep in the truck."

He bit his response before it could get out. *Maybe I'd rather sleep in the fucking truck* was too petulant, and it was goddamn cold out there. Instead, he threw his bag onto the cot between Donner and Masters. Both huddled around Donner's computer watching a DVD.

"That was fast." Benjamin unpacked his sleeping system and rolled it onto the cot. "Nerds."

Masters fixed him with a frank look. "This motherfucker wanna talk about nerds when he the one who played fifteen hours of *Final Fantasy* on one day off."

Benjamin shrugged. "You watched."

"'Cause it's a good fuckin' game. But I'm not the one over here talkin' 'bout 'nerds.'"

He made a good point. The leadership got on them about how often they spent time playing video games or watching movies. But why the hell not? Just like Miller harping on them about smoking, harassing them about what they did to unwind was the height of folly. They could die any day here. Let a guy enjoy the things he liked.

Rather than argue further, he sat next to Donner to watch the movie with them, but his heart wasn't in it. It never was after talking to Tiffany. He should be there, not here. Maybe somewhere higher up the chain of command, someone knew what the hell they were doing. It couldn't just be about money and oil. A lot of people said that, but they were all talk. Things here needed to be fixed. He didn't buy into the "Better over there than in America" logic so many people spouted, but there had to be something.

After talking to her, after thinking about home, it felt like they were wasting time while Dick Cheney made all of his old friends at Halliburton rich. It felt like George Bush playing out some vendetta against Saddam. It felt like the US was trying to corner the oil market.

He looked around at the men sitting around the dusty old tent. At the faded green plastic that had become his newest home. At the wooden floor beneath his feet. He'd kiss the ground when he finally made it back to the United States.

He stood up after watching the movie for only a few minutes. "I'm going to the PX and the showers. Who's coming?"

The missions continued. They searched the houses beside the river for the supposed store of weapons and bombs kept there. It daunted him in a way few other missions did. Strykers had been known to collapse the banks, spilling thirty tons of steel into the cold darkness of the Tigris River. The same river that had seen the birth of civilization became a tomb for any team of soldiers who couldn't escape the back hatch fast enough. If it landed on its top, there was no escape.

Still, it went off without a hitch, as did the ones following it. Benjamin had never imagined a place could be more chaotic than Baghdad, but seeing was believing. The attacks grew more frequent, the area to cover bigger. The other soldiers looked more like soldiers and less like tourists.

He lay on his cot one night, thinking of that very thing while listening to music in the darkness, when the first mortar impacted a short distance away.

Boom!

He removed his earbuds so he could listen better. It had been felt more than heard.

Boom!

Closer. Much closer. In the quiet between rounds, others rose in their bunks. They shared the tent with the tankers, and Silva, the youngest and newest tanker, spoke up first. "That sounds like—"

"Shhh!" Miller held up a hand.

The whistle screamed loud and clear above them, and they all jumped from their bunks to run to the bunkers outside. If you heard the boom and survived, you were fine. If you heard the whistle, it meant you might be next. The enemy wasn't just shooting blindly; they were walking them in.

Boom!

A round landed just outside the T-barriers around the tent.

Boom!

All of them stood in the bunker.

Almost all of them.

"Where's Silva?" Ski asked.

They looked around, even going so far as to peek outside. A moment later, Silva walked into the bunker wearing shoes.

"Oh, fuck," Sutter said. "If you got shoes on, you're wrong."

Indeed, the rest of them stood either half-naked or fully naked. Life and death situations left little time for modesty, not that any of them had much to begin with.

Garcia wasn't having it. "Are you fucking kidding me?"

Silva snapped to parade rest in his underwear, but it was already too late. "Get the fuck out of the bunker and start pushing."

The mortars had stopped, but the pain had just begun for Silva. He pushed long into the night.

The next day, members of first platoon were attached to fourth for a mission. Sergeant Bales and Specialist Caleb sat in the bunk next to Benjamin, who did pushups to pass the time as they waited for the next SP.

"Keep your head up." Bales poked Benjamin with his foot. "I know you're just doing them for fun, but at least do them right."

"Roger, Sergeant." Benjamin kept pushing.

Fitness was a way of life for the infantry. Even now, balls-deep in a war growing more dangerous every day, many of them found time to get to the gym. What the hell else were they going to do? When they couldn't get to the gym, they did what Benjamin did. Pushups, sit-ups, pullups. Burpies, planks, and stretches. Anything to stop the quiet and the calm from getting in. Anything to distract you from the explosions in the night when there was no mission to keep your head in the game.

Everyone lost weight from missed meals, exercise, or just the insane amount of energy it took to keep up the operation tempo. Benjamin had gone to Iraq at one-hundred ninety pounds. He was down to one-sixty. They looked lean. Lethal. The circles around their eyes and the sinew of their muscles made them appear every bit as dangerous as they were.

Bales apparently thought the same thing. "There you go. That looks like a fuckin' killer doing a pushup."

That night's mission went off without a hitch. Someone who had been recruiting for one of the local terrorist networks hid under a bed, but second platoon found him. All the terrorist big talk about being true warriors chosen by Allah, and it never stopped them from hiding or crying. Allah might be working for them upstairs, but he certainly didn't appear to be on their side. Of course, he didn't seem to be on anyone's side in this war.

The next night they prepared for another mission. Rumors flew that they would be going back to BIAP soon. Tidings like that were always welcome.

A dozen feet pounded the dirt outside. Someone, more than one someone now, yelled. Brenner burst into the tent. "Everyone, spin up. We have a mission. Sergeant Miller, Sergeant Garcia, there's a meeting with the first sergeant and the commander." He vanished as quickly as he'd appeared.

Benjamin's heart sank. One mission after the other. One tragedy after the next. If it wasn't a bomb going off and killing a soldier, it was a base being attacked or a high-value target spotted in the most dangerous sections of the battlespace.

"You heard him." Miller grabbed his kit as he walked toward the tent flap. "Nelson, get everyone ready." He and Garcia followed the lieutenant.

Nelson looked as nonplussed as everyone else, but he was nothing if not professional. "You heard him, fuckers. Get your shit on and let's go."

Within moments, they arrived at the vehicle. They performed the same equipment and buddy checks that always preceded a mission.

"Fuck, man." Sutter lit a smoke once everything was finished. "We're never getting out of this fucking area."

Nelson had run off to find Sergeant Miller and get the mission details, leaving Olson in charge. "Just focus on your job."

Olson had been trying hard to live up to his NCO stripes ever since he made corporal. It was the worst rank in the army. Anyone above you saw you as a glorified specialist, and the other specialists saw you as a giant tool that took yourself too seriously. There was no winning.

Sutter took a drag off his smoke. "Thanks for the input, Corporal Olson. I was considering just taking a nap, but your solid fucking advice has shown me the light."

Olson shook his head and climbed into the vehicle. Benjamin couldn't tell if Sutter was trying to be funny or mean. Truth told, both looked alike.

Donner watched the whole thing passively from atop the Stryker. Up there, with the dying light silhouetting him, he looked every bit the soldier he was. Proud. Professional. Pissed off. "I hate this fucking country."

Nelson jogged back a moment later. "A helicopter got shot down south of here in Najaf. We're going down as QRF."

"Are there any survivors?" Sutter asked.

Nelson put his helmet on and climbed into the vehicle. "Nobody knows what the fuck is going on down there. That's why we're going to check it out and recover what we can. We're leaving in five minutes."

The rest of the leadership ran to their vehicles. Within thirty minutes of Brenner's heads-up, they left.

"All right," the recently promoted Major Clinger said over company radio. "Let's move out."

Chapter 16

No one said a word as they drove through the endless desert. To be at work at seven in the morning, they had to leave by five-thirty, a straight drive through the civilized parts of Kuwait and onto the highways. Those highways twisted through a sea of dunes as far as the eye could see. A long, barren road to Camp Virginia. Farther north along the same road would lead them to Iraq, only the camels and coyotes watching them pass.

Jay had wondered aloud on his first day how they kept the road free of sand.

"Street sweepers," his boss, Darrell, said. "Some poor Indian bastard's job is to get out here in a street sweeper every day and brush the sand off the desert."

Jay had laughed, but it wasn't a joke. Even as one of the richest nations in the world, Kuwait had to accommodate its surroundings. He'd been scared it would be hostile, or the language barrier would be too much. They had been ungrounded fears. Few people so much as gave him a dirty glance. Most were kind. Many spoke English. Indeed, the people of Kuwait went out of their way to accommodate strangers.

He hadn't expected that. The last time he'd been here, there had been no real attempt to be civilized.

Kevin, Gini, Darrell, and most certainly Jay weren't morning people, so the drive remained quiet. Out through the sands to the gate where they had to unpack everything. Onto the base and to their station where they switched off with the other crew for forty-eight hours at a time. Ambulance maintenance, then breakfast. There, the conversation started in earnest. In many ways, it wasn't so different from the military.

"So you used to eat like this every day?" Gini asked as she popped a grape into her mouth.

"Sometimes. A lot of times we had to eat MREs."

"I heard those weren't too bad." Darrell took a big bite of sausage.

"I'm sure we can find some if you'd like to try them."

Darrell chuckled and shrugged. "Can't be all that bad. You lived." Unapologetic, stubborn, and no nonsense, he ran their little crew. When Jay had first met the man, he was sure those traits meant it was going to be a long year or two.

He'd been wrong. Darrell, despite being more hardheaded than most, was an upstanding guy. One of the better bosses Jay ever had.

Being around soldiers like this had taken some getting used to. For many, contractors were an oddity. Jay remembered that feeling. Sometimes seeing others who didn't work under the same restrictions was nice; other times it aggravated him that people made a dime off the war. Either way, they were better seen than spoken to.

Being on the other side of that chaffed. No longer one of "us", he was now one of "them." Even when he had the chance to explain what he'd done in the army, it mostly fell on deaf ears and polite nods. He'd quickly given up on it and accepted his role.

After breakfast, they did medical training for an hour and the occasional errand for the Navy TMC[47]. Other than that, their days were clear. Most of their job involved medical work for the military. EMS calls so soldiers could better spend their resources elsewhere. In theory, it was tough work. In practice, they made a call or two a week and enjoyed an obscene amount of free time, much of it spent on the internet or bullshitting with the firefighters who lived next door.

It never failed to impress on him how little his co-workers knew about soldiers, even the few who'd been in the military.

"I almost finished Parajumper school." Brian, one of the firefighters, leaned back in his chair and smiled. "Got knocked out on the last phase. I never deployed or anything, but that was close."

They sat outside and smoked. Jay had to look anywhere but at Brian, lest his face give away what a bullshit statement that was. The

---

[47] Troop Medical Clinic

low, connex-like buildings and tents around them were civilization's rallying cry against the desert. T-barriers and the dirt-filled baskets that had marked the FOBs in Iraq sectioned everything off. But it was different here. Hotter than the devil's balls, for sure, but peaceful. Even as the sun choked the life from the Earth, the barren landscape held a certain beauty.

"That's pretty cool, Brian." Jay flicked his cigarette into the sand outside of the smoking shack in front of the fire station.

"You deployed, didn't you?"

Jay nodded.

"What did you do?"

The way he asked it felt confrontational, even if he didn't mean it that way. That old anger, the sudden burst of rage that put him at a nine just after it passed two, burned in Jay's chest. "I killed fucking terrorists. You know, like a real soldier."

Brian stared, mouth open, as Jay trudged across the sand back to the ambulance bay. It took the better part of an hour to calm down. Even after months there, it was hard to fathom that so many contractors were out of touch with who they were and what they did. They walked around as if they were the stars of the show, not the soldiers. It had everything to do with their job and the money they made. Never mind that even at the best of times, firefighters saved stuff and not people, or that these men never ran a call. Never mind that they looked down on the soldiers they were here to support. In their own minds, they were Special Forces in Afghanistan, not bored EMS in Kuwait.

Everything but the paycheck came loaded with bullshit. He tried not to think on what that said about him.

By the time he'd been there for six months, his friends from the military had been steadily trickling through on their way to the war zones. Nothing drove home how much his life had changed like walking around Camp Arifjan with Staff Sergeant Olson.

"How's Angie?" Jay asked as they sat on a bench outside of a McDonald's.

"She's good, man. The girls are good too. When was the last time you saw them?"

He did the math in his head. "Back in 2007, so four years."

Olson took off his soft cap and rubbed the top of his head. "Can't believe it's been that long. They got big, man. Sprouting up like weeds."

"You going to have more? Thinking of starting up a whole franchise of little Olsons?"

Olson laughed the same delighted giggle he'd used during Jay's entire military career. Something about it made his heart hurt, but he didn't know why.

"I dunno. It's hard to be a dad and a solider, you know?"

He didn't know, but he could imagine. The divorce rate in the military hovered around fifty percent, even as the wars wound down.

Somehow, most of his old team had beaten those odds, but that didn't mean it wasn't hanging over their heads.

The quiet settled in. A familiar quiet like the months they'd spent together in the back of a Stryker when conversation had run dry. Sometimes, even among close friends—perhaps especially— words failed. Sometimes nothing communicated how hard, unfortunate, or strange something was as well as the presence of the people you cared about the most.

Before long, Olson had to catch a plane, and Jay stood with his hands in his pockets, watching his old teammate and friend tread into the lion's den once more.

"Be safe, turd!" Jay waved at Olson as he stepped through the doorway.

Without looking back, Olson waved. You learned early in the infantry that it was best not to look back.

While the chance that a bad call would come down always loomed overhead, Jay wasn't unaccustomed to it. The occasional horrific wreck on the highway popped up, but life in the infantry and then in surgery had taught him to expect the unexpected. It made relaxing hard but action easy.

But the casual way in which contractors disregarded soldiers continued to bother him. The smug comments they made about pay. The outlandish claims about their own contributions to the war. The way they strode around the base as if the war were about them.

161

Jay had missed other things as a soldier as well. Across the Middle East, people from poorer countries were hired to do the work nobody else wanted. Kitchen workers, garbage men, toilet cleaners—anything undesirable was cast down to the "lower class". The US took the cue and did the same. The people who made food in the DFAC[48] were from Nepal. Half the gate guards had travelled from eastern Asian countries. The women who always smiled at everyone as they cleaned toilets came from India. All received well under the minimum wage set by the government that hired them.

When Jay pointed it out, Darrell laughed. "It's still better money than they'd get working in their own shithole countries." He popped a handful of sunflower seeds into his mouth. "Besides, these Arabs treat the TCNs[49] worse. They take their passports and pay them pennies compared to the military. You sure you want to bite the hand that feeds you instead of aiming that anger at the people who deserve it?"

The way Kuwaitis treated TCNs horrified him, and women didn't fare much better. But cultural differences explained that; the US had no excuse. "Are you suggesting that Kuwait and the US are equals?"

"Of course not."

"Then what the hell are you talking about?"

They continued for hours before they both went to bed, and the next day they picked up in the same spot. Darrell never silenced

---

[48] Dining Facility
[49] Third Country National

anyone with his rank or position. He debated religion, life, and politics just as vehemently as Jay.

"You liberals are all the same." Darrell leaned back in his chair, clasping his hands behind his head. "Never mind that the US is one of the most charitable countries on the planet. Ignore that even our poorest have a higher quality of life than the next guy. You always want to pick at something."

The discussion had started with race and gender relations in the US and descended into a diatribe about why liberals hurt the nation. Jay had never considered himself one before, but here he was, taking up for the cause.

"I think being so critical of ourselves is one of our best assets. Our ability to continue asking what's right and wrong, to strive toward making our country better, will always serve us better than blind patriotism. It's what sets us apart from countries where the status quo is just accepted, like Kuwait."

Darrell chuckled, a sound that never failed to sound condescending. "You lefters always say you're making the country better. Obama is on the road to spending more money than Bush, and every time someone points it out, he blames the last president. It's a joke. You want to make the government big and huge, maybe that's what you should be focusing your attention on, not these bullshit social issues. Smaller government is better for everyone. Every new function tacked on makes it unwieldy. It makes it even harder to do the things that need to be done, and it comes with a boatload of handouts that make people lazy."

"Having a government primarily concerned with the welfare of its people is a good thing. I'll never understand why you right-wingers don't see that. We have infinite money for wars while our schools fall apart in Michigan. We can build hundred million-dollar jets, but we balk at food for the hungry."

Darrell laughed again. "I've never heard a guy actually argue for bigger government. Let me guess, you want them to take our guns too?"

"Look out the window at the tanks, bombs, and guns out there. You and I have to control the government with words. People in charge are power hungry, and they have armies. If you think your rifle is going to protect you from them, you're stupid. Didn't do much for Iraq."

Those arguments went around in circles, but it made Jay question where he lay on every issue. For the first time, he realized he didn't agree with the right as much as he'd always thought he did. Looking back, it made sense. Falling into that trap had been easy. First his parents, then the military. Nearly every authority figure he'd ever known had told him at one point or another that the left wanted to destroy their way of life. But Bush had mired them in war, not Obama. The Democrats weren't the ones getting on television and condemning Muslims and Mexicans as if they were the ones responsible for every bad thing happening back home. Democrats strove to build bridges while eight years of Bush had done nothing but burn them.

Once that pebble of doubt fell, the avalanche came. After that, questioning everything else came naturally. Had the war been right? Had his involvement? He'd certainly questioned it then, but that part of his life was a small box marked *Do Not Open 'til Doomsday* that was never, ever touched. He could recall everything then only through a haze. Mission after mission. Side by side with the best soldiers on the planet, the closest family he'd ever had. He hadn't supported the war, but he'd fought in it.

By the time Donner passed through on leave nine months in, Jay didn't want to be there anymore. The money continued to tempt. He never imagined he could make so much and do so little. But with his friends putting their life in danger for a cause, regardless of its morality, the checks became hollow. What was his cause? Cash? Status? There had to be more to life than that.

The first thing he did when Staff Sergeant Donner stepped out of the hangar was hug him. "Hi, boo."

Ever stoic, Donner only stood there with his arms at his side. "What is this?"

Benjamin laughed while soldiers all around them stared. "Emilia told me to do it." Donner's wife hadn't so much asked as demanded.

Donner finally cracked a smile. "I hate you both."

Benjamin backed off and took a good look at his closest friend from the military. A little taller, and a little older, but it was certainly him. "The rank suits you."

He shrugged. "They'll promote anyone who stays in long enough."

"Whatever you say, Sergeant Major. Want to grab a bite to eat?"

They sat in front of the same McDonald's he had with Olson months before, catching up on the last three years. Donner seemed genuinely interested in what the outside world was like.

When Jay finished telling him about his life, Donner raised an eyebrow. "Sounds terrifying."

"You get shot at for a living."

"There's not actually a lot of effort involved in getting shot at."

Jay pulled out a smoke and lit it. "Fair enough."

"How do you know what you want to do? How did you land this job?"

Jay listed off the answers, though the question struck him as strange. He'd never given it much thought. The next step in life had come as a matter of fact after the military. One foot in front of the other taking him down whatever road they found.

Or maybe those steps hadn't come so easy. That unease had been growing for a long time. First in work, then in medic school, and now here in Kuwait. It followed him like a little monster covered in chains, rattling and screaming, always second guessing everything he did.

"You okay?" Donner asked.

"Yeah, I'm fine." But he still changed the subject.

Soon they wandered the base, Jay pointing out the sights. "That's the medic station." He pointed to the shack with two ambulances in front of it. "I work on a different one. For now, anyway."

"Quitting?"

Jay shrugged. "I forgot how much I hate contractors until I became one."

More than once, the squad had sat around griping about the men and women who leveraged their military experience into a job that took advantage of the government's lavish spending. More than one of the old Chargers had done it. He'd respected some of them, and their decision had been a surprise. Some had never been anything more than greedy bastards, and it shocked him less.

"Yeah. I thought it was weird when I heard you'd become one. So what are you going to do?"

He'd been putting a lot of thought into that. He had faced politics, the war, and the world they lived in for what it really was during the year back overseas. More than anything along the way, he'd learned to abhor ignorance. "I'm thinking of getting a degree in history and philosophy."

Donner laughed. "For what? So you can use the diploma to keep you warm when you end up jobless?"

"Something like that. I don't know. I'm beginning to think the only reason I wanted to go to medical school was because everyone told me I was supposed to want a job like that." He clucked his tongue. "I might still do it after. Who knows?"

His experiences in the medical field had been tepid at best, but he'd wanted to make a difference. Plenty of other professions changed the world though. A teacher. A historian. A scholar, political or otherwise. Anything but staying here. Here he did nobody any good.

"Being a civilian sounds complicated. You should join back up."

Jay scoffed. The military had molded the man he was, but the man he was no longer wanted the military. The things they could do were monumental. When they helped those in needed. When they defended the weak. But too many years of bad politicians had soured that. Elected officials took advantage of a good group of men and women and used them for purposes both nefarious and cloaked in shadow. It would be a cold day in hell before he took part in that again.

"I leave that to you and Olson. Besides, from what I hear, there aren't a lot of companies like Charger Company."

"You're goddamn right there aren't. This unit is a donkey show."

"That bad?"

Donner spit. "The leadership is a joke. The soldiers don't give a damn. It's every shitbag we had in Charlie Company but without the advantage of outnumbering them. I never realized what a good unit it was until I left."

Major Clinger and First Sergeant Ferrelo would have beamed if they'd heard that, and Jay said as much.

Donner shrugged. "It's the truth."

They stopped in front of the tent where Donner was staying until his plane left. Jay had to get back to work soon, but it was hard to say goodbye. He'd be seeing Donner on the way back through, but the ever-present "What if?" hung in the back of his mind.

"You take care of yourself." Jay put a hand on Donner's shoulder.

"I will as long as you don't hug me again."

Jay laughed. "Blame your wife for that one. And tell her I said hi when you see her. I love you, man."

Donner cringed. "I love you, too."

They parted ways. Jay didn't see him on the way back. A busy day for EMS threw base-hopping to visit old friends out the window. Gini made an effort for him, but it was no use. The day didn't wind down until well after Donner was on his way back to Afghanistan.

"I'm sorry." Gini laid down in her bunk. "Maybe you'll run into him when he goes home."

Jay didn't bother to tell her he would be long gone by then. The year really had been all about the money, but that would change soon. He'd hit the nail on the head with Donner; something needed to change. He didn't want to be a man only concerned with what was in his bank. He didn't want to continue being angry, and he didn't want the glamourous jobs everyone else did. He'd dreamed of being a writer his whole life, but that could wait. He wanted to make a difference. He wanted to change the world for the better. The military hadn't done that for him then, and contracting wasn't doing it for him now. But maybe going back and getting the education he'd been putting off was the way to get there. Maybe focusing not on what would make him money, but on what would bring him knowledge was the way to proceed.

He didn't know. But as he laid down in his bunk above Gini and closed his eyes to go to sleep, he was excited to find out.

Chapter 17

The drive took longer than he'd expected, and by the time they arrived, night approached. Word had come down that the people in the helicopter were already dead. UAVs[50] and a squad of local MiTT guys on scene had confirmed it. Still, recovering equipment and bodies was as important as saving lives. Not only did the military keep meticulous counts of all their tools, but honoring and preserving dead brothers and sisters was the last respect they were paid. Nobody in the company would shirk that duty, even if they could.

It's too quiet here would be cliché, but it was. Quiet as the grave and just as cold. They had driven through evening's dying light to reach what looked like an empty field. Not that Benjamin saw much of it through Olson's hatch. The sun's final rays peeked over the horizon, soon replaced by the cold January night.

---

[50] Unmanned Aerial Vehicle

"What do you see out there, guys?" Benjamin asked over the headset.

Every time Olson tried to respond, the company radio came alive. Something was wrong. Leadership rumbled about a large group of armed insurgents that had taken down the bird, but they were nowhere to be seen. The helicopter lay where it had crashed. An air of apprehension hung over everything, and the atmosphere tingled as if charged.

"I don't see shit, man. Just a lot of nothin'," Olson finally managed between transmissions.

The drones overhead saw nothing either. At this rate, it would be an average recovery operation. An in-and-out tragedy wherein the most dangerous part was getting to the objective and back. They'd taken up positions around the downed bird, but besides some small-arms fire on the way in, there was nothing.

The next words over the radio silenced the whole Stryker, the whole world. "Charger 6, UAV 2, you have one to two hundred armed insurgents coming towards your positions from the village."

"Did he say a few hundred?" Olson asked.

Nelson kicked him. "Shut up!"

Major Clinger came right back on the radio. "UAV 2, clarify. Did you say one to two hundred?"

"Charger 6, roger. They're coming fast. Over."

Benjamin's heart skipped a beat. A split second of panic preceded a moment of elation. No searching for one bad guy in a whole neighborhood of them. No wondering where the sniper round had come from. No IEDs in the road. The enemies were out there, and they were coming.

"Holy fuck," he said under his breath. "Hundreds?"

The FBCB2 came alive as it was populated with enemy positions. The hair on the back of Benjamin's neck stood up. It didn't look to be a mistake, and it wasn't a wrong transmission. More armed people appeared on the screen than they had seen in the rest of their deployment combined, and they were out for blood. People with good intentions didn't come in numbers, and they didn't come with guns.

Major Clinger rattled off orders at a breakneck pace. Fourth platoon and command set up a perimeter around the downed helicopter while the rest of the company put themselves between the bird and the nearby village. They called for reinforcements, though it would be some time before they arrived. Mortars took up the rear. The top of their vehicle opened, and the 120mm mortar was drawn. No other mortar men had ever fired one at enemies in combat. It had become a running joke in the platoon that they were just a poor man's rifleman. A cheap version of the real infantry. The only thing the mortar ammo accomplished was making them the biggest VBIED[51] in the company.

It felt like a joke, as if some elaborate hoax was going on to make him, all of them, feel like real soldiers and not custodians.

---

[51] Vehicle Borne Improvised Explosive Device

The first shots fired into the last rays of the sun made it clear; this was no joke. The deadly serious business they'd trained for was upon them. More shots followed. Then more. This was it. This was battle.

Olson climbed out of his hatch and into the back with Sutter and Benjamin. "Stay sharp, boys. It looks like we might actually get to do our job."

Sutter rolled his eyes but said nothing. Outside, the darkness spoke with gunfire. From their position, they could see nothing but rocky ground and trees. To their left the other mortars set up the 60mm for closer engagement. The rest of the company dug in, dismounting from vehicles and setting up on the ground, ready to catch any enemies that marched toward them.

But whatever action had found them was impossible to see from where Benjamin stood in the back. He craned his neck looking over the top of the vehicle and into the night behind him, but they were too far back for it to be of any use. The company and platoon radio called targets and updates on air support and reinforcements. The FO[52] radioed to make sure they were set up and ready to go. The goddamn battle of their lifetimes was happening right outside, and Benjamin could see nothing.

"What the fuck, man?" Benjamin poked his head through the open top again, looking for some sign of the enemy he knew was

---

[52] Forward Observer: Soldiers that mark and calculate targets for indirect fire and support. When referred to as "the FO" it's often remarking on the officer in charge of the FO team.

there. He saw only the green and black landscape of night vision. "I can't see shit."

"Shut the fuck up, Benjamin!" Sutter punched him in the leg. "The adults are trying to fucking work."

His heart pounded as if he was a child caught somewhere he shouldn't be. "Hey, fuck you. I'm an actual infantryman, and I'm goddamn excited."

"We're all excited, you dumb fuck. But why don't you—"

"Both of you shut the fuck up!" Nelson yelled. "Talk over the radio one more time, and I'm going to fuck you up."

"All right, look sharp," Clinger said on company net. "Keep your heads down and your eyes up. There's a lot of them."

Close air support buzzed by overhead. Specter Gunships flew in. Out there, his friends fought an enemy he couldn't see. It infuriated him. He continued peeking over the top of the Stryker, looking in the direction of the battle.

"Keep your fucking head down!" Duarte threw a rock at him from the entrenched position next to their vehicle.

"Fuck." Benjamin sat back down in the seat, waiting for the call, the signal to rain destruction down on someone.

Sutter hit him again. "Calm the fuck down, asshole. Goddamn. You're a soldier, not a child."

Benjamin shook his head. "Next person who hits me is getting a fucking butt-stroke to the chin."

Gunfire continued outside, but Sutter stared daggers at Benjamin through his NODs. "Yeah, *Ben-ja-min*? You gonna hit me?"

Benjamin kicked him hard in the shin. "Your move, faggot."

Sutter laughed, a dark and threatening sound. He set aside his rifle, but Olson jumped between them. "Both of you settle the fuck down, or I'm gonna fuck both of you up!"

"Look out," Donner said over the headset. "Olson's going to get you!"

"You shut the fuck up too, Donner!" Olson threw a magazine into the hellhole separating the driver's seat from the rest of the vehicle. He looked as though he wanted to crawl in there and get him, but his kit was too big.

Nelson had been furiously going over the data on his computer, preparing to fire should they need to, but he finally turned his attention to them with a scowl. "All of you shut the fuck up!" Nelson punched Olson in the helmet. "We're fighting a war, you bunch of fucking children. Benjamin! Man the 240. Be ready to come down quick if we have to fire." He returned to his computer.

Benjamin grunted and did as he was told, shooting Sutter one final, fatal glance. The tension crushed them all, as heavy as stones. They were so close to doing their job, so close to fulfilling their purpose, but it wasn't happening. Just like every other day of his career, something exciting was happening just over there. Just around the next goddamn corner. He'd sit in this damn Stryker with

Olson, Donner, and Sutter, and all of them would continue to be as useless going into the next day as they had been on the previous one.

"And make sure you stay low in the hatch, stupid." Nelson punched him in the kneepad, making his knee buckle. "Don't get shot because you're a dumbass."

"Roger."

The cold bit through his anger. So easy to forget the winter when the adrenaline set in. But forces of nature never let you ignore them for long, and even through an IBA[53], a helmet, and two shirts, the chill nipped him. The enemy could sneak up behind them, but with a wall of bullets and men between them and here, it didn't look likely. So he stared out into the darkness in front of the vehicle while the battle of a lifetime continued behind him.

The FO and the CO worked with the Specter and other air support as they rained death and destruction on the enemy. Behind them, a mini-gun buzzed again and again. Explosions as loud as any IED he'd ever felt rocked the landscape. Then, the first thousand-pound JDAM[54] dropped.

The CO keyed into the radio. "Get ready. Keep your heads down."

The whistle, that damn whistle, rang out overhead, and then the world shook. No exaggeration at all—the force of the explosion literally rocked the ground under their vehicle. Benjamin turned to watch in awe as fire glowed in the distance, a black pillar of smoke

---

[53] Interceptor Body Armor
[54] Joint Direct Attack Munition

rising to the heavens where the ground had been churned as if by the fury of God.

"Jesus fucking Christ," Olson said. "That's insane."

The FO came over the radio. "Mortars, get ready. Sending numbers to you soon."

Benjamin's heart skipped a beat. "We're firing?"

"Looks like it." Nelson pulled up his FDC[55] information on the small computer in front of him. "Get back here and start prepping rounds with Sutter."

Donner moved the vehicle slightly to get into position. Everything had to be just right to drop a bomb into a ditch or onto a building from half a mile or more away. Complex math went into angle and direction based on the FO's information. The amount of explosive propellant left on every round determined velocity, a nightmare of calculations that he thankfully didn't have to worry about.

He'd barely cut open the first round when Nelson yelled, "One round, HE quick. Deflection! One-eight-two-two. Elevation zero-nine-four-three."

Olson repeated the commands as he and Sutter adjusted the cannon.

Benjamin's hands shook, and he fumbled with the chargers on the round. He tore off his gloves just as Nelson called out the command he never thought he'd hear. "Hang it!"

---

[55] Fire Direction Control

Sutter took the round from him and held it over the top of the mortar.

"Fire!"

He let go, and the round slid down the tube, making the metal-on-metal *tssst* he'd heard so many times in training. They ducked down, moving away from the top of the weapon, as the round reached the bottom. When it hit, the pin struck the primer, and with a deafening *boom*, it launched toward the enemy.

He grinned as the round fired off into the darkness, and laughed aloud when it detonated ten seconds later. Olson had a smile plastered across his face. Sutter held up a hand for a high-five, which Benjamin gladly gave him.

"Keep your head in the game, morons!" Nelson yelled.

More commands came in. Donner adjusted the truck slightly, the cannon's angle changed as Nelson screamed numbers and the gun crew repeated them. "One round, HE quick! Hang it! Fire!"

Another round roared off into the night. Benjamin laughed as he watched it go. A maniac sound, foreign even to his ears. Soon the smiles faded, his along with the others.

"Six rounds, HE! Fire for effect!"

And they did. They rained hell on their foes. The 60mm mortar next to them did as well. Another JDAM dropped. The earth shook once more, and the pillar of smoke grew higher and higher. A whole grid square[56] on fire behind him. A village lay in that direction, a

---

[56] 1 Square Kilometer

place where people had once laughed, loved, and died. Now they only died. Nothing could be left there.

A burning, angry thing in his chest he'd never known before pulsed with joy at the thought. These people had tried to kill his friends, his family. They had come at his brothers with guns, trying to take them the same way they had Sergeant Henkes. Their death brought him a bitter, ugly joy, and that little ember glowed brighter.

They ran three gunships black on ammo[57]. They dropped as many mortars as he ever had in training. Two JDAMS exploded and countless ammo spent. At some point in the night, Bravo Company helped close off any vectors of escape from the destruction, denying their foes any hope.

As the sun rose over the edge of the Stryker, Benjamin and Sutter shivered in the cold of the back. The heater did little to aid them when the top was open, but they could do nothing for it. Being a soldier meant being uncomfortable more often than not. And now more than ever they were soldiers. More than any other men he'd ever met, they were warriors. They were killers. They had become the people they'd always known they would be.

The battle had long since stopped. When the last Specter had run out of ammo, the enemy had ceased advancing. Charlie and Bravo Company had wiped out most of them, according to the radio chatter. Not a hundred, hundreds.

---

[57] All rounds expended

Miller pounded on the back hatch. "Benjamin! Sutter! Get out here. They need help getting people to the CCP[58]."

Sutter jerked awake. "Roger, Sergeant."

They opened the hatch and jumped out, scanning outward for any sign of foes. When they found none, they followed Miller toward the village.

He expected the scene of destruction that lay in front of him as they crested the hill. Low, square buildings Iraqis loved so much, many ruined or scorched with fire. Roofs caved in. Walls collapsed. Nothing but stone rubble remaining despite the building next to it being intact. Soldiers milled all around, working. Pulling security. Sergeants and officers barking commands and creating order from chaos.

What he hadn't expected were the bodies in the street. Blackened from fire, their limbs drawn up as often as not. One had his hands over him still, as if that could have protected him from the destruction. Soldiers checked for signs of life on those who were clearly dead. They even poked at bodies barely recognizable as human. They looked like the remains found at Pompeii.

But not all the soldiers looked for life. Some celebrated death. Rosenberg shook a water bottle in front of a man's face as he crawled across the dirt, a trail of blood behind him.

"Water... Water..." He inched toward the men as Rosenberg laughed. "Please, mister." His intestines spilled out of his side,

---

[58] Casualty Collection Point

trailing in the dust behind him and turning muddy brown where the sand and the blood mixed.

Just as he entered arm's reach, Rosenberg threw the bottle. "Get the fuck out of here, you terrorist piece of shit."

A strange feeling burned next to that angry ember in Benjamin's chest, the same as the day he'd wheeled the man into the TMC after Henkes had died. Not pity, but he couldn't place it. They deserved what they'd received. These assholes had tried to kill his friends. It wasn't regret, or guilt, it was something else.

They passed a building filled with dozens, hundreds, of charred bodies. Hands up. Faces burned into grimaces of pain, as if they'd known before the end what a horrible mistake they'd made. A gunshot fired inside. Benjamin, Miller, and Sutter raised their rifles, but a moment later Sergeant Griffin and his team walked out. Benjamin could only describe the expressions on their faces as accomplished. They looked like men who had seen the face of God. Eyes lit, faces reverent even when grim.

A group of people—men, women, and children, many of them wounded, gathered outside another building.

Miller shook his head, and his face hardened into a mask of disgust. "They were bringing ammo and weapons to the men shooting at us last night. Should put a bullet in all of them."

But that was the difference between the enemy and them. They didn't cut heads off on television and attack innocent people to drum up support for their cause.

The moans and screams of the wounded filled the building they used as a CCP. One woman tried handing her child to Doc Lars. "My baby! My baby!" The child's skin was burnt black below the waist. Its abdomen held a bloody wound the size of a grown man's fist. It had long since breathed its last.

That same hot sting of anger Miller had expressed woke in Benjamin's chest. She was responsible for her child being dead, or her husband, or whatever moron had led them here. Play stupid games; win stupid prizes. They didn't get to assault the most powerful military on the planet and cry when it didn't go their way. The United States could be out of this shithole country faster if people like her weren't attacking them at every turn, giving them further reasons to stay.

"I need to get back to the mortars. You two do what Doc says, clear?"

Both of them responded in unison. "Roger, Sergeant."

Miller walked back the way they'd come.

The riflemen had already cleared the village of any remaining fighters. Only the survivors remained, praying to Allah they could get through the day.

Looking out at the destruction, he finally realized what he felt. As he stared down at the bodies of the men killed in that village, the truth became clear. The only difference between them and his brothers-in-arms was a circumstance of birth. The deciding factor in whether a man became a terrorist or a soldier came down to the place where your parents shit you out. No patriotism mattered. No religion made its stamp on a man before his creation. Dumb luck ruled the day. Like real estate, enemies concerned themselves with location, location, location. These people were just in the wrong place at the wrong time. That's why he couldn't hate them, why he couldn't hurt them even when he could get away with it. It wasn't pity, and it wasn't just anger; it was acceptance.

That didn't excuse them, and it didn't illicit any sympathy that hadn't been there before, but it did ease that ember inside of him, if only a little. They had done their jobs. They had crushed their enemies in the most decisive way possible, and they hadn't lost a single man while doing it. Nobody in the world could have done it better. Countless generations of warriors had culminated in this battle, in these soldiers, and they had won.

"Hey, guys." Doc Lars threw a box of blue gloves at them. "Put these on and give me a hand."

Chapter 18

His hair had been short when he left, but when he stepped off the plane in Chattanooga, it was nearly to his shoulders. His beard had grown ragged. All the free time he'd had in Kuwait had resulted in a lot of time in the gym, and his muscles bulged under his shirt.

Allen waved him down as he stepped out of the gate, running over with a smile on his face. He looked Jay up and down as he approached. "You look like a homeless caveman."

"Ooga booga, bitch." Jay threw his arms around him.

Allen embraced him in return before holding him out at arm's length. "I forgot you turn brown in the sun. Careful. Someone around here might figure out you're actually Mexican."

Jay laughed.

The plane had stopped in Germany and then in New York before landing in Chattanooga. It had been exhausting. Too wired to sleep, he had read a book, watched a movie, anything to try and contain his excitement at being home. When the plane had finally touched down, the sun just peeked over the horizon. This was it. He was back.

Outside the airport, he stopped and took it all in. The green. The fresh air. The heat from humidity and not just the sun mercilessly beating everything down. He'd remembered all that, but he'd forgotten how good it was to come home after being away.

"I love this place." He jogged to catch up with Allen.

"Hell yeah, man. We missed you! I got the keys to your apartment for you." He handed him an envelope. "The rest of the guys are down to help you move your stuff out of storage tomorrow."

They drove down the highway and over the Tennessee River toward the apartment. It was beautiful. Boats filled the July water. People played and splashed on the shore. So different from the place he'd spent the last year. The money he'd made would get him through college, supplementing what remained of his GI Bill and ensuring he could live in comfort. Even so, he wondered what he'd been thinking leaving a paradise like this.

"I'm staying here forever. When I'm dead, just toss me in the river."

Allen patted him on the shoulder. "If we throw you in the river, you'll end up leaving."

"Fair point. Just dump my body somewhere nice. Burn it to a crisp and throw it at people on the street."

They laughed all the way back to the apartment, a nice place in the heart of the Northshore area. Close to the university and everything he could want access to. Another slice of heaven.

"You do the honors, man."

He fit the key into the lock and opened the door.

Jay nearly had a heart attack. All his friends waited inside. "Surprise!" they yelled.

Laughing again, he dropped his bag and hugged them each in turn. They all spoke at once, either to him or to each other. They had brought food and beer, the latter illegal in Kuwait. He'd never been a fan of it, but surrounded by good company, nothing had ever tasted as sweet as that first sip.

It was a day of new beginnings and renewed purpose, of putting the past in its place. He'd gone off searching for something that had been here all along. He was right where he belonged.

He was home.

Chapter 19

They took care of the people they'd just rained hell on. Had warriors of the past done that? Did Spartans cross the battlefield after a slaughter, offering water to their enemies? Did they provide succor to the foes that had tried to kill them the night before? That separated the refined from the barbaric, the ability to set aside their hate and do what was right. And yes—maybe some, like Quinn, would rather they glassed the whole place. But even he did what was morally right when the time came.

Or maybe they could see that gray area because it wasn't their home being destroyed.

Whatever the case, they stayed on site through the day, pulling guard, aiding the wounded, and speaking with awe about the accomplishment now under their belts. Few were the men privileged to see their warrior training pay off in such a spectacular way. A tense night of battle became a jubilant day of near celebration, patting one another on the back for a job well done and a battle well fought.

The first sergeant announced over company radio as they pulled into the FOB the next morning that the chow hall would stay open late for them. Everyone in the Stryker cheered. War made a man hungry. They parked just outside the DFAC, and a hundred thirty-odd killers made their way inside, laughing, cheering, and roaring at one another and everyone they encountered.

Nelson, Benjamin, and Sutter walked in just behind Howard, who pointed at his boots as a young woman stared. "See that blood? That's the blood of my fucking enemies."

Benjamin laughed, and Nelson rolled his eyes. The young woman picked up her tray and made a quick exit.

Indeed, they were filthy, bloody, and the bags under their eyes hung low. Benjamin had been turned away from the chow hall for being too dirty in the past, which struck him as ludicrous. Even here at war, someone stood outside and checked IDs, telling those who spent their time doing dangerous work that they needed to change their shirt if they were too sweaty.

No such person guarded the door today, and if there had been, the men of Charger Company would have run them over. They yelled as they ate, laughing and joking with one another in the secret language only a soldier could understand. Far from telling them to be quiet or reminding them they were professionals as leadership normally would, many of them joined in.

But before long, even the thrill of battle and the lure of hot food couldn't keep them awake. Slowly, two by two and three by three, they left the chow hall to head back to the tents that were their temporary homes.

Soon, only Donner and Benjamin remained. Nelson rose from the table and took his tray with him. "Be back soon, you two. You need to get some sleep. There's going to be a lot of shit to do today."

Then they were alone, the DFAC nearly empty. The TCNs stood behind the counter and watched them in a way that always screamed unease to Benjamin. Not fear. They weren't afraid; it just seemed they didn't understand the cultures they'd attached themselves to, either American or soldiers.

He stared at his empty tray, lost in thought, when someone stopped next to their table. He ignored them, though it did pull him from his pondering. You learned early in the military not to volunteer in any way if you wanted to avoid work, and he most certainly wanted to do that now.

"I never thought I'd live to see the day that Jay Benjamin ignored me."

Benjamin looked up, shocked by the voice he hadn't heard in years. He'd know that Maryland accent anywhere. Kennedy stood in front of him, sergeant rank on his chest and a Ranger tab above his airborne tab and unit patch.

"Holy shit!" Benjamin jumped up from the table and hugged him. "Man! I never expected to see you here!"

"Never know who you're gonna see. How the hell are ya?"

Kennedy was the only reason Benjamin was still in the army. Realizing that perhaps the military wasn't as fun or glamorous as it had seemed from the outside, Benjamin had done anything he could to get out in basic training. Again and again, the leadership let it slide, infuriating him. Eventually, they'd had enough.

His basic training first sergeant had slammed the paperwork on the table. Kennedy, the platoon leader, had watched the whole thing, standing to the side with the drill sergeants.

"You're done, Benjamin," the first sergeant yelled. "My army doesn't want a piece of shit like you. Understand? Who's responsible for this asshole in the platoon?"

Kennedy stepped forward. A college specialist. A soldiers' soldier, even then. "I am, First Sergeant."

Kennedy had gotten it up and down. *He* was responsible for the actions of those under him, and he needed to learn that. *He* should have tried harder to turn a young soldier around, despite being one himself. *He* had more rank and therefore more responsibility. Benjamin had never been more ashamed in his life. This man had done everything he could to help. He'd never said an unkind word. He'd never lost his temper. He'd never judged Benjamin solely on his merit as a soldier when everyone else had. Kennedy didn't deserve the ass-chewing he received.

Later that night, the drill sergeants had both of them in their office.

"I have never seen such a disregard for the welfare of your fellow soldiers. I cannot believe this is what my army has come to." Drill Sergeant Gallentte wouldn't even look him in the eyes.

That had been a powerful moment. Benjamin's father had once told him there were a handful of turning points in life. Every day certainly mattered. The road from who you were to who you would be shifted all the time. But the big ones, the things that stood out, really set you on a path.

Benjamin had always wondered what those moments would look like. Freshly eighteen, uneducated, and scared, he'd never seen one before. But Kennedy, only a few years older, knew the score. He looked proud. Disappointed. Not upset like everyone else. Maybe it was because he'd tried again and again to take Benjamin under his wing when he'd given the man no reason to do so. Even when everyone else sneered at him or told him he was worthless. And even after getting yelled at for it, after catching shit for looking after someone who didn't care enough to do it for himself, he didn't look angry.

"I'm sorry," Benjamin said.

Drill Sergeant Gallentte looked up. "Excuse me?"

"I'm sorry, Drill Sergeant."

The room went cold. The other two drill sergeants in the room stared daggers at Benjamin. Kennedy retained his composure, looking straight ahead at the wall.

192

"Why?" Drill Sergeant Gallentte leaned forward on his desk. "And why the fuck should I believe that when you've done nothing but give everyone a hard time?"

He broke his rigid parade rest. What could he say at a time like that? That reality had come crashing down when he'd seen what his life would be like in the army? That he was exactly the scared teenager he appeared to be? That he wanted to be better?

"I've been acting stupid, Drill Sergeant. But if you give me one more chance, I promise you, I'll give it my all."

The three drill sergeants eyed one another for a long, silent moment. At length, Drill Sergeant West spoke up. "You've missed too much training. Even if the first sergeant let you stay, you'd have to get recycled back to day one."

The thought of starting all over with new people six weeks in sent his mind reeling, but Kennedy saved him from it. "Drill Sergeant, I can take responsibility for Private Benjamin. I can help him catch up on the training he's missed."

All three men turned to him before exchanging silent words with each other through eyes alone. Years later, Benjamin would understand the silent language of men who'd seen much and done more.

Drill Sergeant Gallentte folded his hands on his desk. "You hear that, Benjamin? You have other people picking up your slack already. Is that what you want?"

There was no good answer to that question, but he tried to find one anyway. "I want another chance to be a soldier, Drill Sergeant. That's all."

And that had been that. They'd gotten the platoon together, because in the army, everything affected the community. The drill sergeants told them Benjamin would be staying, and that it was their responsibility to help him. Most had sneered, or watched Benjamin with hard, cold eyes. The kind only a young, bellicose soldier had, and only when they're eager to prove how strong they are. But Kennedy, true to his word, had helped. He'd rallied the others around the cause.

"A chain is only as strong as its weakest link, guys. Remember that."

A few months later, they all graduated together.

The last time they'd been in a room together, he'd asked Kennedy why he'd stuck up for him.

"Never take responsibility for something you wouldn't be proud of, Benjamin. You're a good soldier, don't forget it."

Kennedy went on to do whatever things a super soldier did, and Benjamin went to Fort Lewis and joined one of the most accomplished units in the United States military. It was all just a roll of the dice.

But here he stood like a ghost from the past, smiling at Benjamin in a DFAC on some nowhere base in Iraq. Benjamin shook his head. "I'm good, man. Great!" He flicked the Ranger tab on Kennedy's arm. "I see things are going pretty smooth for you too, Sergeant."

Kennedy, ever humble, only smiled. "I've been lucky to have good leadership and a great unit." He looked Benjamin up and down. "Looks like you landed in one yourself."

Benjamin glanced down at his uniform, covered in a day of dirt, grime, and blood. In garrison, a clean uniform meant a good soldier, or at least one who knew how to play the game. In combat, that wasn't always the case. "We just got done with something big, man. You wouldn't believe me if I told you about it."

Kennedy tilted his head. "You were part of what just happened in Najaf?"

Benjamin did his best not to smile. "How did you hear?"

"I work directly for my brigade commander."

This time, he didn't hold back. A grin spread from ear to ear. "Shit. You're going to be a seven in seven and Sergeant Major of the army before you're forty, you bastard."

Kennedy shrugged. "I don't know about all of that. I'm just doing what I can." He checked his watch. "But I have to run. I'm already a little behind for a meeting. It was great to see you, brother." He stuck out a hand, which Benjamin shook.

But Benjamin couldn't leave it at that. This man had done more for him than almost anyone else in his life. He'd taught Benjamin the value he had, even when he couldn't see it himself. That deserved more than a handshake.

"I never said thank you for what you did for me back at Fort Benning, man."

Kennedy waved it away. "You don't have to."

"No, I do. I wouldn't be the man I am today without you. Thanks for that, brother. From the bottom of my heart, thank you."

"I'd do it all again. You're a good man, Benjamin. Don't let anyone tell you otherwise." They held one another's gaze before shaking hands once again. "You be safe out there. I want to hear about all the best-sellers you write when you get back home."

Benjamin laughed. He'd forgotten he had told Kennedy about his writerly aspirations. "I will, man. You do the same."

He walked off, and Benjamin watched him go until he left the chow hall. Never in a million years would he have imagined running into Kennedy. The military was certainly a small place, and he thought of the man often enough, but it was unexpected. Unexpected, but right. Charlie Company had crushed their enemies, and he'd found the man responsible for putting him into a position to do so. More than luck, it felt like fate.

Donner set his spoon full of cereal down. "Jesus, man. Why don't you blow him?"

Nelson had been correct. Checking every piece of gear, maintenance, layouts, and meeting after meeting about not calling their families to brag filled the next day.

The Commander hammered it home in a formation that afternoon. "Seriously, guys. Don't call your mom. Don't call your dad. Don't call your girl or your Uncle Jim-Bob and tell them about this shit. The news will come out soon enough. If any of you fuckers break OPSEC[59], you're going to have both of us and the battalion CO standing in front of the brigade CO. As soon as they're done ruining your career, I'll ruin your day. Don't do it."

But after formation, on the way to the phones, Jared had smiled at him. "I'm going to break the shit outta OPSEC."

Benjamin laughed, but he wouldn't follow suit. He didn't have long in the military after they got back, and he didn't want to jeopardize that. Paranoid or no, every day held a new and exciting journey to see how the army would fuck you.

But when he reached Tiffany on the phone, she already knew more than he had guessed. "What happened? The news is going crazy here. They're saying soldiers killed a whole bunch of civilians."

He thought back to all the people they'd detained in Najaf, and to the mother holding her dead baby. He still didn't pity them, but the memory rose of its own accord. "There weren't a whole bunch of civilians; there were a whole bunch of people shooting at us."

Silence. "You were there?"

"I can't talk about it over the phone."

---

[59] Operational Security

But she wouldn't be deterred. "They said soldiers got killed. Was it anyone you know?"

"No. Everyone's fine. They were other soldiers. It was a big fight, but I really can't talk about it on the phone. They monitor these lines. I had to sign paperwork saying I wouldn't spill the beans." The old World War II propaganda photos flashed through his head. *Loose lips sink ships!*

He expected her to drop it, but instead, she grew angry. "Why? People here have a right to know. The news is making everyone paranoid."

He wanted to tell her that if the news caused the problems, she should take it up with them. Media had no place in a war. Combat was dark, ugly, and bloody. The people back home couldn't fathom that if they'd never been a part of it, but they were all armchair generals. They all had some input about how nice or mean soldiers should be. Once a war began, kindness was a hindrance, not a help. If the people back home really wanted this war to end, they would shut off the camera and let the rough men work.

"Sorry, angel," he said. "I really can't talk about it."

She finally let it go, but it occupied his thoughts even after he hung up. He mentioned it to Jared on the walk back to the tents.

Jared reacted about as expected. "The news is saying we killed a bunch of unarmed people?"

Benjamin nodded.

"Fucking assholes."

"Right? What the hell do they think is going on over here?"

Jared shrugged. "It's fucking war. People back home are pussies."

He told Nelson and the rest of the mortars about it when he returned to the tent.

"That's stupid," Sutter said as he lay in his bunk. "Those morons have no fucking clue what goes on here."

It was eerie how those sentiments mirrored his. Or maybe they'd spent so much time together that it was only natural they share a similar mind.

Nelson didn't stop reading his book, not even bothering to look up. "Fuck 'em. Just do your job, and don't worry about what the peanut gallery has to say."

Masters looked up from doing pushups in the corner. "That fine-ass girlfriend of yours said that?"

Benjamin nodded.

"It's 'cause she needs a dick in her mouth."

Benjamin threw an MRE at Masters while the others laughed.

A week later, they returned to BIAP, the closest thing they had to a home. Second closest, anyway. Nothing could ever feel as familiar as the Stryker they spent most of their army lives in. It certainly lacked in comfort, but nowhere felt safer. Still, those comforts greeted them when they returned. It all felt familiar, but nothing was the same. How could it be? They had stood upon mountains of enemies and come out not only unscathed but laughing. Every ounce of pain that had taken them to this point was measured and found to be enough.

They'd proven they were greatest warriors on the planet, and they knew it.

Chapter 20

Back in the army, people laughed at the idea of college being hard. Jay had joined in just as much, making fun of the teenagers who bitched about finals while people died on the other side of the world. But as he grew older, he realized how silly that was. Someone else's struggles not being as literal as your own didn't diminish theirs, and having a hard time didn't mean everyone else had it easy. Misery wasn't a competition, just a fact of life.

In truth, college wasn't hard. At least, not in the ways he thought it would be. Showing up and doing the work came easy. Knowing you had years to go before you finished and years after that before you were where you wanted to be; that was the hard part. Maybe it was the entitlement so many people claimed of millennials, or maybe he thought he was owed more after serving his country. Either way, it took its toll.

Just as it always did, his enthusiasm drained, but this time he noticed the pattern. Go somewhere new; find that thrill. Work on it for a while, trying to make it last. And when it didn't, he gave up. He got down. But this time would be different. This time he was going to stick it out and make something of it. And he liked school. The infectious open-mindedness of it made it easy to stay involved. Every day he met new people, learned new things. How people could go their whole lives without something like that was beyond him. Education was a drug, and he wanted more. He joined the Honor's Program at the university in his second semester.

Harman, the religious studies professor, became his favorite teacher. More than that, really—a mentor. They spent a lot of time together, eating lunch, discussing the finer points of religious studies and where someone could go to continue them after getting a bachelor's.

He noticed the strains as the second year wore into the third. One day, they sat in his small, cramped office filled with books and religious icons from around the world, discussing an upcoming term paper. Jay had always imagined a professor's office would look like that. Harman had travelled widely in his youth and had at least some insight into almost every culture or spirituality.

Now, however, Harman eyed him with concern. "Are you all right? You haven't looked so good lately, and I've noticed your work… How do I say this? It isn't as imaginative as it used to be."

A polite way of saying he was slipping. "I'm fine. Just a rough semester." Here he at least felt like he was making some progress in his life, but there should be more. The puzzle lacked a vital piece, and if he could only find it, he would be content.

Harman leaned back in his chair. An old man, older than Jay had imagined when he'd first met him. His vitality and enthusiasm hid it well, but he was deep into his sixties. "If you're getting burned out, there's no shame in backing off a bit. You don't have to do everything all at once. And you've been through more than most students around here."

Every life told a story; some were just more exciting than others. "Honestly? I don't know what it is. I just feel…" Incomplete? Upset? He'd had to stop watching the news because it made him angry. Every time someone mentioned Iraq, his temper went from a one to a ten in a heartbeat. "Off."

Another veteran in a sociology class had worked up the nerve to quote MacArthur on the subject of conflict. "Who more desires peace than a man who has seen war?" The others in the class, including the teacher, nodded as if he'd said something profound.

Maybe that had been true of total war. Battlefields torn asunder. Millions dead. But for the winning side in a horribly unbalanced conflict, war became glamorous. People loved you when you came home, and that made it easier to love yourself. A little closer to that dream of celebrity. A little bit more Madonna, a little less poor kid from the ghetto. But millions had died, and for the side that lost, it wasn't glamorous and proud at all. Jay had kept the thought to himself, but it bothered him for days.

Harman picked up on all of it. "Like I said, don't be shy about relaxing. Everyone needs some time to themselves."

Jay left the office with promises to do just that, but taking a break felt no different from quitting. With so much left to be done and so far to go, who could stop before they reached the end of whatever road they were on? He knew it was ridiculous even as he thought it, but that didn't make it any easier to shake off.

Things grew worse that afternoon when he went home.

Someone had made a Facebook group for Charger Company, just a page where the old dogs of war could keep track of one another, bullshitting about current events and popping in when the nostalgia bug bit.

But grim news filled the page today. Someone had killed sixteen people just off the base where Charlie Company currently lived in Afghanistan. Rumors flew. All the questions were the same.

"Was it one of us?"

"Are any of our guys hurt?"

"Was it Bales? I heard it was Bales."

Major Clinger had finally gotten on and told everyone to wait. The news would come out soon enough.

Jay's heart dropped as he stared at the screen. You expected bad news when so many of your friends were soldiers. Those men with the folded flags were always on the way to the house of a widow you knew. But this? The news said one of theirs had killed children, that he'd lost his mind and slaughtered half a village by himself. It was too outrageous to be true. That sort of thing happened in Vietnam, not now. Not with the honorable sons of bitches he called brothers.

He stared at the screen, his stomach turning over. They'd proved they were better than this a thousand times during their deployment. While terrorists cut the heads off civilians and raped children, they gave water to people on their patrols. After Najaf, they hadn't wandered the village killing what remained; they'd provided medical aid to those in need.

He shut off his computer and stared at the wall. It wasn't about him. He said it repeatedly in his head, but that didn't make him believe it. That was family, and what one of them did, or how one of them felt, affected them all.

He finally lay down to sleep that night, his rest fitful and his dreams chasing him down endless corridors. Dark thoughts followed him through rooms he'd seen in Iraq as countless familiar faces stared back at him. All his friends joked in the back of their Strykers once more, but he was no longer one of them. Just a civilian, a man who didn't belong. They knew it. He knew it. When he woke the next morning, he turned on his computer, dreading what he might see.

It was all true. Robert Bales had gone on a bloody killing spree.

He skipped school that day, instead sitting on his couch and scouring every new piece of posted news. He couldn't look away. His heart hammered as if he'd been there when it happened, even though he'd been sleeping comfortably in his bed on the other side of the planet. He kept waiting for the retraction, for the change of names, anything that would remove a brother from this awful story. It didn't happen.

US Soldier in Afghanistan Kills Sixteen Women and Children.
Sixteen Dead at the Hands of US Soldier.
Soldier Goes on Killing Spree in Afghanistan.

Every new article posted on every news site only confirmed the truth. The Charger Facebook page buzzed with activity. Some comments offered stoic shows of support in a hard time, but some attempted at defending him. They made it sound as if he were a victim because of his time overseas, because of a country that didn't care or leadership that had let him down.

One of their old lieutenants posted some trite nonsense about the state of the country they lived in. "We live in a nation that pats us on the back with one hand and slaps a toxic waste symbol on with the other." As if US citizens were somehow responsible for Bales' monstrous actions. What he really meant was you couldn't teach a boy to kill and then be shocked when he does. Again, and again, and again. A million different ways. But that was bullshit. They weren't children, irresponsible for their actions. They were soldiers.

A bubbling anger rose inside Jay the more he read. These weren't the actions of a cold, hard professional; they were those of a murderer.

It bled into weeks as Bales' story came out. A man condemned to four deployments in rapid succession. The stress of combat. The danger of PTSD. But anyone who'd been in the military knew there were ways out. Combat stress, buddy checks, going to the docs and telling them you needed an exit. The people defending him tried to absolve him of murder, because that was exactly what he was. A murderer. The only thing preventing a round commendation of his crime was the color of his victims.

There was no escape from news of the war. Jay went to school as normal. He worked as normal. He lived his day to day life unaffected, but not really. Sitting in the library after class one day, staring at a book he wasn't reading, he realized he'd been hiding from Iraq since he left. But between Bales and the rumblings on the news that the US had pulled out of Iraq too early, it was impossible to hide anymore. The commentary became ubiquitous. They were using the face of a brother to describe exactly how things had gone wrong. They were using his family to hammer home what an unjust set of wars they were, especially in Iraq, where no WMDs or nuclear materials had ever been found. Not only were they stomping all over the name and "honor" of a man he'd once known, they were questioning everything he'd ever accomplished. They were questioning the family he came from.

Jay returned home that night and took a long, hot shower, staring at the tile walls as if they held some answer. He tried to convince himself it didn't matter. He knew what they'd done. They weren't monsters or murderers, even if some had become so afterward. They'd been there to make a difference, not because they had a vendetta against the people they fought.

But how many off-color or outright hostile remarks had they made about Islam during their time there? About all the locals? Hadn't that terp told him when they were stuck at that shithole outpost before Najaf that Iraq had been better before they showed up? Hadn't Quinn wanted to turn the whole place to glass?

His heart beat so hard that he thought he was having a heart attack. His breath came in ragged gasps. He couldn't breathe. The walls closed in around him. The water grew too hot, and the steam choked him. He shut it off and sat down in the tub, squeezing his eyes closed as the room spun. After all they'd done, after all he'd accomplished, he was going to die right there in the shower, and nobody would find his body. Nobody to care that he'd passed.

He skipped school for two straight days, expecting the worst. Was it a heart condition? Cancer? Would he find out he had only a few months to live?

But the doctor at the clinic up the road explained that it was nothing. "I know it feels like there's something wrong, but I can promise you there isn't. Panic attacks are normal, and they can be very alarming."

Jay stared at the man in disbelief. It hadn't been his imagination. He hadn't been able to breathe.

His skepticism must have been clear on his face, because the doctor continued as if he'd said it aloud, "Are there any new stressors in your life? New job? Recent breakup?"

Jay shook his head, but it was a lie. What did you call it when you began to believe that your friends might have died for nothing? That perhaps some of the brothers you'd once thought of as the very models of honor and decency weren't?

"Well, think about it. Next time it happens, just take some deep breaths. Relax. Do whatever you normally do to calm yourself. And remember, they're transitory. They'll go away after a while."

But the next day they didn't. Or the day after that. As one week slowly wore into the next, his nerves frayed, and every day became a game. Holding his breath until something set him off. Waiting for the world to pull the pin that would see him explode.

Chapter 21

Back in Baghdad, things had fallen into their old routine. When they weren't working, slaving away under an enemy's scope for all they knew, they prepared to do so. No resting. No reprieve from the constant pressure. Progress became a concept, not an accomplishment. Quite the opposite. More attacks every day. More IEDs. More random gunshots from people only half-seen, if at all.

"Just remember, guys. Small steps." Major Clinger said it repeatedly, as if it had any meaning. And then a new commander, Fellers, replaced him.

During the change of command, the new CO spoke as if they were already family. "I'm looking forward to working with you all. I've seen the work you guys are doing from HHC, and I'm not going to lie, I'm excited. You're the best damn unit in the brigade, and I'm going to live up to it."

But Benjamin eyed him warily, first from formation and later on missions. You didn't break up a winning team halfway through the game.

Still, things changed daily. If the military excelled at one thing, it was shaking up the status quo. They pulled Donner out of the driver's hatch and replaced him with Benjamin. He had to get ground-level experience, according to Miller.

"If you wanted to stay out of the hatch, you should have reenlisted. We want people who are staying with the team on the ground, not short-timers looking forward to getting out."

Benjamin's temper flared as Miller spoke to him from the bunk across the tent during the squad meeting. Being in the hatch didn't bother him. Even if it meant he had to put extra hours into the vehicle when it came to maintenance, it also meant he'd be able to ass out during the boring missions where the mortars spent half the night staring at a wall.

It was the tone. They'd become real warriors, and Miller still spoke to all of them as if they were children. Miller even treated Nelson, who was the same rank, like a moron who'd only wandered into the uniform.

But beyond that, Miller treated him as if he was on the outside already. As if by saying he didn't want to do this anymore, he was a civilian just playing at being a soldier.

"Do you have anything to add, Benjamin?" Miller stared at him through those Coke-bottle glasses.

"No, Sergeant. I'm good."

And so he drove them around, mission after mission. It had been an easy transition, and it kept him out of the line of fire. If they wanted to treat him like a short-timer, so be it. Others had warned him before deployment that as soon as you made it clear you wanted out, you were stuck in the *other* category. Not quite civilian and not really a soldier anymore, just the guy who'd get picked last for kickball from now on. You'd be the one shoved into the driver's hatch usually reserved for new privates even though you'd been in three years.

He tried to let it slide, made all the easier because he had no control over it anyway. Everyone was on edge. Just lions roaming their cages, from privates to the goddamn colonels. Roaring at one another, marking their territory.

Every day grated his nerves. The EFPs terrified everyone most of all. They could blast straight through a Stryker, killing anyone unfortunate enough to be hit regardless of where they were in the vehicle. They hung water cans filled with mud from the slat armor on the sides of the vehicle, some genius from brigade claiming it would slow the projectiles enough to prevent it from piercing the armor. Anyone who had seen what they could do to a tank knew mud and plastic weren't going to slow shit.

By some miracle, no EFPs hit anyone in the company, but it was close. Near misses occurred more often than was comfortable. Detonations right behind a vehicle or in front of one. Something blowing up only ten feet from the rear air guard of a Stryker and cutting a telephone pole in half across the road.

They all knew what kind of soldiers they were. Better than most, better even than their peers in the battalion. But more than that, they knew they weren't invincible. Making it through whatever was left of this deployment without any more casualties was a matter of luck, not ability.

And rumors came down every day on that front too. During a squad meeting, Benjamin had asked Miller point-blank if he thought they would be extended.

"Does it fucking matter, Benjamin?" Miller slammed his notebook on the table. "If you get extended, you do your fucking job until they tell you you're done. Is there a fucking problem with that?"

"Negative, Sergeant. I was just curious."

"And you don't think I would tell you if I knew?"

Benjamin did his best not to sigh with irritation. "Roger, Sergeant. I'm just tired."

Miller stared for a moment longer before his expression softened. It always did. As antagonistic as he could be, he always caught himself afterward. "We're all tired, man. Just do your work. They'll let us know what's coming down."

Once the meeting ended, Miller and Nelson left to do more of the endless paperwork the military loved so much, leaving Olson and Duarte in charge.

"That guy's a real prick," Simmons said, lying down in his bunk.

Masters took off his boots and did the same. "Yup."

Duarte had been with Miller longer than anyone but Nelson. He went back and forth on whether to support or condemn the guy. This time, it looked to be the former. "It's tough leading troops. Cut him some slack."

Benjamin rolled his eyes as he sat down to write in his big green notebook, but he didn't let Duarte see it. Miller shielded his guys from going out of the wire as much as possible. He looked out for them as if they were all his little brothers. But that was part of the problem; they weren't his kids, they were soldiers, and he shouldn't treat one like the other.

They continued their missions, sweating every day, lugging guns and gear across a desert made tame. Big news eventually came down the pipe, but it wasn't the news they expected.

Miller came back from his meeting to find all of them lying in their bunks. "Pack your shit up." He slapped Benjamin on the boot as he walked by. "All of it except your cots. We're moving."

Everyone sat up, but Duarte was the first to question. "Like, moving tents?"

"Moving to a different FOB. Taji. Our mission has changed. We're going to be supporting the 82nd Airborne in and around Sadr City."

Sadr City was the Baghdad slums, and no coalition forces had been in there in years. One guy, Muqtada al-Sadr, supposedly ran the whole place. It was more complicated than that, but that was Benjamin's understanding.

So once again, being good at their jobs had bitten them in the ass. "When are we leaving?" Duarte jumped off his bunk.

"Three days."

Everyone groaned, but they got to work too. It never ceased to amaze Benjamin how different the military was from the expectations. From the outside, it appeared to be a perfectly oiled machine. Every edge crisp, every formation practiced, and every action decisive. To watch the movies and hear the stories from family members, it looked as though they could mobilize a million soldiers in a week's time. In theory, they could.

Up close, the cracks in the system showed. Everything was a last-minute rush job. The standards were continually blurred to accommodate the situation, and it was never an easy affair. Professional, maybe, but it always came with more stress and pain than necessary if the higher-ups had only shared their plans with everyone else a little sooner. But they didn't live in that world. The missions stopped, and soon they were packed and ready to go. Well-oiled machine or no, they excelled at what they did, even when jumping through their asses at the last minute to get it done.

The result ended up every bit as bullshit as the effort. If Benjamin had thought the tents were bad, the airplane hangar where they stuck the entire company on Taji was worse. Being right next to a burning garbage pit and only half a mile from a bunch of tanks destroyed with depleted uranium rounds made it that much better.

They parked their Strykers in the big, empty lot next to the building. It had clearly seen better days, as had the rest of the FOB. It was an artillery base, and even then, in the middle of the day, rounds popped off in the distance. Just another barren wasteland in a country full of them. Dry, cracked earth as far as the eye could see outside their new home, only T-barriers and berms to mark the edges of the nothingness.

"Oh my fucking god." Simmons jumped out of the Stryker and joined Benjamin in looking out over it all. "This place is a fucking shit hole."

Donner sighed as he climbed out of the vehicle. "Welcome home, Simmons."

Benjamin lit up a cigarette. It might have been funny a few months ago, but nobody laughed now. The big green weenie had struck again. He grabbed his bag off the truck and stepped inside.

It wasn't much better than the outside. Bunk beds from front to back. The whole company in one long, narrow building. Nothing even approaching privacy. They'd gone from two-man rooms to twelve-man rooms, and now they'd come full circle to everyone in one place. With a bit of bad luck, they'd all be sleeping outside before April.

Nelson slapped him in the back of the head as he walked by. "Don't look so down, buttercup. You can sleep next to me."

Benjamin followed him to the small corner set aside for the mortars. "I'm going to sleep inside of you as soon as you close your eyes."

Quinn and the rest of his squad laughed as they walked by. "I always knew you mortars were a bunch of fruits," he said as he tossed his things onto a nearby bunk.

It didn't take long before they were getting ready for the missions that would begin soon. The only shining light in this dank hole was the supply office being far enough away that they wouldn't get many details. Unfortunately, that meant the chow hall, PX, and phone centers were far away too. When Miller overheard Sutter make that exact complaint, he shook his head and sighed loudly.

"You fuckers don't know how good you have it, that's the problem. You're at war, not at your fucking mom's house back on the block. Be lucky you have a bed. Thank fucking God you have phones, and you can call whatever little piece of tail you have back home waiting for you."

He'd set his things down next to Nelson's, which meant they were all sleeping in a ten-by-ten foot area. Stuck sleeping so close to a guy who made his eighteen-hour work days miserable was the biggest bummer of all, but he kept it to himself as he lay down that night.

The nearby artillery firing at all hours didn't do much to keep him awake, but its boom chased him through dreams. Explosions rocked the roof in his unconscious mind, mortars breaking through the flimsy wood that provided no real protection. He and his one hundred thirty-man family, wiped out in the space of a few heartbeats.

He found little rest that night, and when he woke, he worried it was a portent of things to come.

In addition to having their own area to cover, they also ran most of the missions for the local 82nd battalion. A few things came with the territory in the military. You did PT. You slept late on the weekends and short on the weekdays, at least in garrison. More than anything, the greatest truth of the army was that airborne soldiers thought their shit didn't stink.

He'd felt bad for them when he pulled into the little base where they were stationed on the edge of Sadr City. Nothing more than a four-story shopping center converted to a base, surrounded on all sides by tall buildings hiding unseen enemies. Rocket and mortar attacks occurred often, and small arms fire more so. A whole battalion along with their support staff stuck in a place the size of an office building.

Most of that pity evaporated when he saw them work. Or, as the case was, when he saw them never work. Their AO[60] was too

---

[60] Area of Operations

dangerous, and as often as the Strykers covered down on patrols, he almost never saw or heard of any of them in the battle space. But every time they returned to the little outpost with the White Falcons, as their unit was called, the soldiers strode around as if they were top dogs.

"More like Blue Falcons," Nelson said as they pulled in one day. "Bunch of buddy fuckers."

One of their companies had been relegated to a nearby outpost that was nothing more than a tent surrounded by T-barriers with high sight all around it. Those infantrymen acted like soldiers, and they treated Charlie Company with the same respect they received.

"Those guys back at the outpost are fucking pussies," a specialist said to Benjamin while they smoked outside the tent one day. "They barely patrol. They don't do shit. But they all talk a big game."

Benjamin kept silent and listened. It was one thing to joke with the rest of the mortars or to laugh when one of their own disparaged another unit, but talking shit about their unit to their face wasn't done. Military bearing separated the good soldiers from the great ones.

"They hardly even bring us hot chow out here."

Considering the constant danger at the outpost, it was a bummer to hear. While Benjamin and the rest of the crew pulled gate guard in the Stryker one day, a man had walked toward them with a food cart. The road leading to the entrance was long and surrounded by buildings, leaving only a narrow alley straight to the gate. Another "tactical oversight" in a war full of little goofs by supposedly smart people.

The man pointed the cart down the alley, opened the front to reveal an RPG[61], and launched it at them.

He vanished before they could return fire, and a rocket, several pounds of explosive with their names all over it, screamed at them as it flew down the alley.

"Get down!" Nelson yelled.

Benjamin sat in the driver's hatch, safe as houses, but he flinched as it soared toward him. At the last second, it veered off, as RPGs often did. Instead of hitting the Stryker, it sailed past the gate and blew up against a nearby T-barrier, doing little more than surface damage.

But Benjamin's breathing stopped as if it had killed one of his own. He'd watched it coming right at him, and even knowing it couldn't possibly hurt him inside the Stryker, his insides scrunched up, bracing for the impact that would kill him.

"Are you all right, Benjamin?" Nelson poked his head into the hellhole behind Benjamin.

He turned to Nelson, hoping the fear didn't show on his face. "Yeah, I'm fine. Everyone okay back there?"

Nelson nodded before standing up and announcing the explosion over platoon radio.

---

[61] Rocket Propelled Grenade

It only took a smoke and some water to calm Benjamin, but the thought that this other company dealt with it daily daunted him. He might think the rest of the Blue Falcons were a joke, but those men were the real deal.

Unfortunately, that only scratched the surface of the misery. Every day someone shot at them or chucked grenades at the Stryker. Every night they pulled missions in the most dangerous parts of Baghdad to combat terrorists. Supposedly, the local militia had allied with the US. They had standing orders not to shoot anyone with a black and red head dress, members of al-Sadr's militia. In theory, they were currently friendly with the US forces cleaning up his neighborhood.

In practice, you could never tell who might try to kill you. Men dressed as women to get by checkpoints. Parents strapped bombs to their children and sent them into crowded markets. Did the higher-ups really think wearing a certain kind of headband meant someone was off limits? Anyone outside of a US uniform was just as likely to kill you as the next as far as Benjamin was concerned.

He made that comment to Miller one day before a mission.

Miller spun on him, his brows furrowed and his jaw clenched. "You think you know better than everyone, don't you, Benjamin? Well go ahead, *driver*, shoot one of them and see if I don't have you in a military prison so fast your head fucking spins." He stormed away before Benjamin could respond.

221

A response burned on the tip of his tongue. But why bother? It wouldn't fix anything. They would go out and get shot at the next day, same as the last. Stirring the shit pot wouldn't make him feel better, and it wouldn't send them home. Miller had always been a coward, more afraid of getting in trouble than willing to get the mission done.

But what was the mission now? To end terrorism in Iraq?

Benjamin only shook his head as Miller walked away, but Sergeant Rosenberg had overheard the whole thing. He patted Benjamin on the knee from where he sat on a nearby bunk. "That guy's a douche. Don't sweat it. All the line platoons are working on the same ROE[62] as you are, buddy; if it shoots, shoot back."

Plenty of that happened in the battlespace. It was only a matter of time before someone was hurt.

It didn't take long, but by some miracle, it wasn't a member of Charlie Company. When third platoon came back from a mission one day covered in blood, Benjamin ripped out his earbuds and jumped off his bunk, kicking up dust from the wood floor.

"What happened?" he asked Rosenberg as he walked by. "Is everyone all right?"

Rosy looked tired. The bags hanging under his eyes hung lower than usual. He forced a smile and slapped Benjamin on the shoulder. "We're all fine, buddy. Rocket attack on the outpost. A few of the Blue Falcon guys got killed, a bunch more hurt."

---

[62] Rules of Engagement

Their medic, Lars, walked by. Blood coated his arms up to his elbows and his legs to his knees. Unlike Rosenberg, he looked mad. He kicked a box full of paperwork that flew across the bay. Nobody said a word to him.

Rosy watched the whole thing with Benjamin. "Don't worry about him. He'll be fine."

The rest of the story trickled out over the next few days. The rockets had torn through the building, killing half a dozen people. One man survived a direct hit when one had blown through a wall next to him, but it had cut off both legs. By some miracle, it hadn't exploded. What a world they lived in when a thing like that was a blessing.

Benjamin told himself repeatedly as the weeks rolled on with the 82nd that such was the life of an infantryman. If he'd wanted something easy, he would have been a cook. Rumors of extension didn't make it any easier to swallow. Every day new ones came down, and Miller had gone from saying they would wait and see to warning that they should expect it.

"Don't go calling home about it or anything," Miller said during a squad meeting. Once, those meetings had seemed intimate, or at least personal. The closest of brothers standing around discussing their dark work. Now people played video games in the background, or yelled at one another as they walked by. "Nothing is official. I just want you guys to be prepared mentally for whatever might come down."

They talked about it on the way back from a mission one morning, a company-level event in Sadr City to find some bomb maker that had killed a few of the Blue Falcons. It had gone nowhere, as usual.

"If we get extended, I'm going to combat stress," Simmons said. "I'm going to throw down my rifle like Santero and just fucking walk on."

Santero had been a rifleman in third platoon who'd quit only a few months into the deployment. He'd threatened to kill himself if they didn't send him home. When his squad leader had called him on it, he put the rifle in his mouth and finger on the trigger. Only a timely tackling by Rosenberg had prevented a mess in the tent.

"You do that and I'll let you shoot yourself, turd," Nelson said.

Benjamin laughed. "I'll console your sister if you go out like that, Simmons. She'll need a man in her life to be her rock after her brother offs himself."

Simmons snorted. It took a lot to dampen spirits on the way home from a mission. "If she needs a real man, she should really look somewhere else."

Everyone cracked up. As stressful as it was pulling twelve- to eighteen-hour days every day, as bad as missions could be, the promise of hot food and a warm bunk did a lot to soothe a soldier, if only for a while.

Nelson sighed into the mic. "I think both of you—"

A gunshot cracked outside the truck. Nelson silenced at once. After a moment of quiet, Benjamin opened his mouth to ask if they saw anything, but company radio came alive.

"We've got a guy hit; it's—"

That had been third platoon, but they cut out before they said a name.

"Oh, god." Benjamin's heart sank. It had only been a matter of time. He fucking knew it.

Nobody in the truck responded, and a split-second later, the first sergeant came over the radio. "Put up a perimeter. I'm taking the MEV[63] and pulling up."

Radio chat picked up as the company faced outward to defend the tradeoff. Strykers pulled off, staggering the direction they faced. A few hundred tons of military hardware, nearly a hundred fifty of the meanest killers on the planet, and everything went silent as they waited on the side of the dusty road to hear news of their wounded brother. The radio buzzed with chatter, but one of the big rules was never to use names over comms.

"All right, we got him on the truck. We're closer to the Green Zone. First platoon, get us there."

The CO squawked in after the first platoon PL responded. "Charger Seven, Charger Six. I need an initial and a last four. Over."

"Charger Six, Charger Seven. Golf. 6783. Over."

---

[63] Medical Evacuation Vehicle

"Who was on that truck? It was 3-2, right? Who was on there?" Simmons was shaken; how could he not be?

Only the vehicle's beeps and the roar of the engine filled the void. It took Benjamin a moment to figure it out. His mind worked slowly. He'd never been there before when one of his had been shot. "Griffin. Sergeant Griffin was on there, and he would have been out a hatch."

Reality set in. They would have said so right away if he weren't hurt. Only a few minutes had passed, plenty of time to get on the radio and say, "Just a scratch! Just nicked him!" They didn't put someone on the MEV and rush them to the Green Zone if they were fine.

Benjamin's thoughts rambled as the radios deafened. "I'm sure he's fine," Benjamin said aloud. "We were moving when it hit. I'm sure he's okay."

Nobody spoke. Not an insult about being naïve, not an affirmation, and not a denial. His heart hammered, and he was sweating despite the mild weather. It wasn't supposed to go like this. They'd be home soon. There wouldn't be an extension, and the people still alive would all be leaving, no goddamn exceptions.

He sped along as fast as the vehicle in front of him would allow, knowing the thought was foolish. Knowing it was too late.

Chapter 22

"So what? We tell our kids that violence doesn't solve anything and then blow each other to hell without a hint of irony? Where's the sense? Where's the reason?"

The crowded bar hushed as the two men confronted one another. Jay had known it was stupid before he did it, but he couldn't help himself.

The young man in front of him was clearly military. Marine, if he had to guess. A crew-cut and dog tags under a crisp, white shirt gave it away. More than that, it was how a military guy carried himself. The chest puffed out. The pride in each step. Not knowing if someone was watching but still walking as if in a parade. Jay wondered if he still walked like that.

"Violence solves more problems than anything. Read a goddamn book." He spit the words at Jay and took his shot off the counter, downing it in one gesture. His friends stood behind, clearly not part of the crowd around them.

Jay didn't know if this asshole was about to start a fight or not, but his blood pressure shot up regardless. He shouldn't have come out by himself to a crowded club. Too much stimulation. Too much wondering if someone was going to take a swing. "Yeah, you fucking Neanderthals are really fond of saying that. I have read a book. I've read a few. You know what else they say? Violence starts more problems than anything else. What a fucking coincidence."

The young Marine locked eyes with him. He didn't know if the kid would swing, but he wanted him to. His blood pumped a thousand miles an hour in his veins, and he wanted to take it out on someone.

"Whatever you say, man." The Marine flicked a salute at the bartender and walked away without looking back, his friends in tow.

Jay paid his tab, slamming his card down on the counter. He could see his pulse in the veins on his wrist. He couldn't remember the last time he'd been angrier.

To pretend that violence would solve the issues now popping up in Iraq was the height of folly. Violence had caused those issues. It was plain to see that he and his brothers had contributed as much to it as any terrorist in the country.

Nobody seemed to have a grasp on where ISIS had come from, or at least not the assholes on TV talking about it. He'd done his research as best he could, knowing it was stupid. Knowing it would only stress him out at a time when he didn't need it. His grades slipped every month, his boss had noticed his change in attitude, and even his friends remarked how he'd shifted in demeanor over the last year. How he'd become angrier, more distant.

And none of it mattered anyway. As much as pundits tried to defend all the Western meddling in the Middle East, it was responsible. You only had to trace the lines back far enough. Not that the locals were free of blame. The perceived differences and resulting violence between two factions of the same religion was just as culpable as the interventions from Russia, the US, and other world governments over the last hundred years.

Either way, the situation remained the same. All the hard work that him and his friends had put in, all the pontificating about building a better world, had led up to this—the destruction of the country they'd tried to save.

Some people thought the Iraq government and military would pull through, but he didn't. Daesh broke open prisons and overran bases. The region had descended into chaos, and too few of their neighbors were willing to help. Everyone threw out blame and shined their own politics, too busy to unite and do something about the problem. The response of the American people seemed to be an overwhelming, "Who cares?"

But he did, and the rest of his old teammates did as well. He walked through the crowd of young, happy people. Beautiful specimens of the human condition living free and uninhibited, blissfully unaware or uncaring that on the other side of the planet, people died by the hundreds.

He saw the young Marine on the way out. The man waved, and his friends didn't look the least bit angry. But in Jay's head, they were rabid monsters, members of the same group of people getting on TV and defending the situation overseas. Halfway through the long ride home, he realized he'd been the asshole. He'd been the one listening to a conversation he wasn't a part of. They hadn't shared their thoughts with him, but he'd shared his, blurting it at them as if he would get some sense of satisfaction regardless whether they agreed or not.

The Scenic City passed by out his window. Chattanooga. A beautiful, Southern town as far removed from that anger and violence overseas as one could be. It had rained earlier in the day, and in the mid-fall darkness, the whole city shone. The light from the houses on the mountain and the glow of the enormous buildings downtown lit the way behind him, reflecting off the river as he crossed it. Once, he hadn't been able to stop looking. It was so different from the place he'd grown up. Even the nice suburbs of Detroit looked ugly in comparison.

But somewhere along the way, he'd stopped noticing. His thoughts turned inward and ran through his head over and over, a steel ball trapped in a perfect cylinder. Nothing to slow it down. Picking up speed with every rotation.

He lit a cigarette and rolled the window down, letting in the fresh air. Instead of going home, he drove around. Sometimes it helped get his temper back under control. It cleared his mind to cruise the back streets of the city with the music all the way up and the glass all the way down.

Not tonight. He rolled around for an hour before giving up, going back home to stare at the ceiling in his bedroom. It took hours before sleep found him.

He stared out the third-floor window of UTC's language building. It wasn't a great view, but it was a nice day. A little ways down the quad, he could see the Confederate cemetery. It had just been an interesting tourist spot for the years he'd been in town, but of late, he wondered about the men buried there. How many regretted what they'd done before they died? How many had been lied to or realized too late that they didn't believe as strongly in the cause they fought for as they first thought? Did the survivors go on afterwards, broken and wondering what they'd accomplished?

Did a thought like that mean he was broken too?

"Well, Benjamin?" his Roman history teacher, Professor Corvo, asked.

"Hmm?" Jay looked up to find Corvo and half the class staring at him. "I wasn't listening."

Several of the younger students giggled, but Corvo was not amused. "I'm aware, that's why I called on you. Try to focus."

He did try for at least part of the hour. Not only did he like the subject but also the class. Corvo came from snob stock front to back, but he was a good teacher, and his energy made it hard not to enjoy his antics. Jay almost felt bad for caring as little as he did.

Almost.

It was his last class of the day. Rather than go home and study like he needed to, he jumped in his car and cruised around more. The thoughts about the soldiers wouldn't leave him alone. Everyone spoke about past generations of the dead as if their logic, morality, and reasoning were beyond reproach, especially in the United States. The South, as horrible as it had been to black people, still received the benefit of the doubt. People called them heroes. Some people even tried to argue that it had never really been about slavery.

But he knew better than to believe those men hadn't thought the same way people did now. They must have doubted, especially after the war. Those who lived long enough, past the South's destruction and decay through to its eventual resurgence, must have seen the way the wind blew. They must have realized they were on the wrong side of history, both by the books and through morality.

Would he feel that way too?

He could see it at times like those. Endless years of humanity and their wars stretched out before him like an ocean. Never still. Always trying to be heard. Sometimes nothing more than crashing waves and others a monsoon. A hurricane. A towering thing of water and wrath so big and inscrutable that it seemed as if it would swallow the world. And then, nothing. Just the lapping waves once more. That voice in the chaos trying to be heard.

And what came after? When the water was calm again and the need for warriors had passed?

"Man, if the military went back, I'd re-enlist in a second," Simmons had once remarked.

Whether or not it was talk didn't matter. They all felt some responsibility to the place where they'd fought and bled.

Jay's brother had made a similar remark the month before, when Jay had been visiting his family in Michigan. "It's a fucking shame what's happening over there."

Their father had sold both of them on the military, pimping his children out to Uncle Sam from the time they were old enough to hold G.I. Joes to the time they signed on the dotted line. Jay watched his younger siblings run around the crowded living room, playing with the dogs. He wondered if Dad had sold his youngest children on it the same way.

"You could have shared how much the military sucked before I joined up."

Theo shrugged and took a swig of his beer. "I would have, but by the time I realized it, I was in Iraq."

Dad had certainly made it sound more glamorous than it was. It was irresponsible of adults to send children off to war, and as much as teenagers would argue otherwise, they were children. What would Dad have done if he or Theo came back in a pine box? If they'd been killed like so many of their friends had on a back desert road for nothing more than a daily patrol and the promise of a brighter future?

But they hadn't, of course. They came back different. Maybe a little dented, but alive. He'd been tempted to air those same thoughts out to Dad, but he already knew what the man would say, or close enough. He wouldn't use the word unpatriotic, but he'd imply it. He'd justify the war by saying we had to keep the fight in some far off and barely imaginable *over there*. Both of those ideas had been so persuasive when he was a child, but he was a child no longer.

He pulled into his apartment and shut off the car, staring at the building as he drifted through days past and events gone by. Putting all the blame on Dad wasn't fair. The recruiters had been just as willing to sell horse shit, and the people of the United States were ever willing to worship a hero, no matter what he might be under the uniform.

But one thing stuck out more than any other. It had been his responsibility. He had been a child, but he'd signed on the dotted line. And he'd swallowed every hook they'd sold. He and his had been heroes, warriors out of legend who were going to change the world. Someone had once chimed that PTSD was knowing you would never be as cool as you were when you were a soldier. That the most interesting thing about you had happened when you were just a kid, too stupid to realize the significance of the world-changing events happening around you. Happening because of you.

He'd been willing to die for his country. Many had. They'd said it, after all, some aloud and some quietly. That they'd give it all. Generals and presidents could talk a good game about making the other guy die for theirs, but they didn't live in the belly of the beast, and the enemy had no country to fight for. They might have wanted to survive deep down, but they wouldn't, and they hadn't. Part of them had died over there, and for some, the rest followed when they returned home. Sometimes it didn't take long, and sometimes they held on for years, gasping for breath as a piece of themselves rotted away. At the heart of all of it, they'd been the ones to make the choice; they'd pulled the triggers, and the final blame for anything that happened after that rested with them.

He laughed at the morbid thought as he stepped from his car. If was just that after all. It wasn't fair to reduce everything they had been to nothing more than a willingness to die, especially when so many of them had come home. They'd been willing to die for one another more than a cause, and if anything, that should have been a good thought. There had once been one hundred thirty men in his life who would die for him.

Still, it was easy to wax sentimental and let it go when it wasn't his country being destroyed by the latest, greatest terrorist threat. Forgiving yourself was easy when you came home as a loved and respected member of the privileged few who'd been to war. Meanwhile, somewhere far away, people died by the hundreds. Lives shattered and ended. Millions displaced. There was no laughing it off for them, no way to make it okay with logic or reasoning.

His own laughter died on his lips as he let himself into his apartment and did his best to forget.

Chapter 23

He sat on the back ramp, smoking a cigarette, surrounded by the best friends he'd ever had. For once, even Miller didn't ride him about killing himself with lung cancer.

The Green Zone buzzed with activity. They'd parked in a little gravel lot Benjamin couldn't find again with a map and a bird's eye view. They seldom came out to the Green Zone. This was a peaceful area. They didn't need grunts coming in and ruining all their fun. They had MPs to keep a lid on things.

They'd raced inside and parked while the MEV and third platoon rushed Griff to the hospital. Someone in the company had shot at a car that had gotten too close, as if they were still out in the shit and not sitting in the eye of the storm. Nobody said a word over the radio. There were more pressing concerns, even for the most professional of them.

He'd forgotten it was his birthday until someone had mentioned the date on the way back. Twenty-one years old. Sitting and wondering if his brother would live when most people would be out drinking themselves stupid only made him angry, so he tried not to think at all. But other thoughts crossed his mind unbidden. The young man who'd watched their Stryker on Christmas had only been a few years younger than him. Fifteen or sixteen. People would call it barbaric to send such a young man to war back home, but Benjamin hadn't been able to buy beer until fifteen hours before. The irony wasn't lost on him.

Nelson picked up a handful of rocks and began throwing them one by one at Simmons, who sat a short distance away on the ground. Each one pegged him in the back of the head, but Simmons continued smoking, ignoring it.

"Don't lose your focus, guys," Nelson said. "Keep your head on a swivel and remember where we are."

That advice could have come straight out of Miller. Benjamin pulled his gaze from the ground and watched outward. One of the top rules of war: watch what the fuck was going on around you. So easy to forget when something like this happened. If it had been a war where people were dying all around them, he supposed he would have grown used to it. But it wasn't. They'd only lost two members of the company—

He huffed and stood up, pacing back and forth. Griff wasn't dead. Wounded, sure, but that didn't mean dead.

"Chill, dude," Sutter said.

Benjamin ignored him.

238

Despite the early date, the heat had already set in. Not terribly, but within another month or two it would be unbearable. During their last deployment, before Benjamin had arrived at the unit, they'd made snowmen in the winter. Iraq was a country of nothing but extremes. Saints and sinners. Snow and fire. Battle, and the endless waiting in between.

A Humvee pulled up in front of the Strykers. The newcomers left their vehicle and approached the mortars, the closest crew. The CO and first sergeant had gone to the hospital to see to Griff.

It felt wrong to have interlopers on this private moment. They were worried about a friend, and that shouldn't be shared.

An Air Force MP—security forces, as they called them—walked over to the group. Another man and a woman stepped out of the vehicle behind him and approached. All wore deadly serious expressions. "Who's in charge here?"

Everyone around the Strykers stared at them. Nelson stepped forward, but the man spoke again before he could get a word in. "Someone shot at a civilian vehicle from this company. I'm going to need everyone's weapon, and I need to see whoever's in charge."

The absurdity of it all struck Benjamin like a blow, and he laughed so hard he choked on his cigarette smoke. "Hear that, guys? He's gonna take our guns."

Benjamin didn't know jack about Air Force ranks, but he guessed by the look on the man's face that he thought he was important. He placed a hand on the pistol at his waist. "Do you think this is a joke?"

By now, everyone nearby looked that direction. Sutter stood up on the ramp and cradled his weapon at the low ready. "You're out of your fucking mind if you think you're getting anyone's weapon."

The man's eyes widened in anger.

Nelson jumped in before anyone could say anything else. "Woah! Nobody is taking anybody's weapon. What's the problem?"

The XO appeared from the crowd of angry infantrymen standing around and observing. "What's the issue, Sergeant Nelson?"

Benjamin watched the exchange. Apparently, someone inside the alleged car was as important as this loud Air Force turd, at least to themselves. Suddenly, everyone had amnesia. Had there been a car? Did anyone shoot? No way, not in this company. Trigger discipline is our watch word!

The Air Force man insisted they continue the game, but nobody else was playing. He took the company information and left.

What the fuck did he expect? Did he honestly think he was going to walk up to a hundred thirty trained killers and take the only thing between them and a bloody death on some terrorist's beheading broadcast? He'd lost his fucking mind.

As he walked away, white-hot anger replaced the mirth Benjamin had felt at the absurdity of the situation. One of them lay bleeding somewhere nearby, maybe dying, and the military concerned itself more with some car getting too close to an infantry convoy. In a war where people bought big-screen TVs and took tours of the sights around BIAP, nobody cared about actual soldiers but soldiers. Not these useless POGs, not the politicians back home, and certainly not the people on either side of the political debate.

They were on their own. Happy fucking birthday indeed.

The service started at seven that night. All across the hangar where the company lived, a somber mood reigned. Missions had continued the past few days as if nothing was wrong, as if one of them wasn't dead. Time marched on.

Benjamin had barely said a word all day. He sat on his bunk, staring at the blinking cursor on his screen. At times like this, he should write. Do anything to get all that black ooze out from inside of him. But the words wouldn't come. Not a month before, he'd sworn up and down he was going to keep a journal of the war, but he'd stopped after only three days. He didn't have the heart for it. He didn't want to relive every stupid event only hours after having done it the first time.

And so he stared at the blinking cursor until Nelson walked up and slapped him in the leg. "Time to go."

They walked together a few hundred meters down to the hangar that passed for a chapel. When Henkes had died, it had been a bigger affair, but things were different then. The missions had grown harder, more dangerous. The brigade had been spread out, and the man from the support battalion who played taps on the bagpipes whenever a teammate died was nowhere to be found. Just as well. Benjamin didn't feel up to hiding his tears or jumping at a twenty-one gun salute.

The infantry almost never wore emotions on the sleeve. These were hard men, and hard men had no place for that. Still, as they trod down the dusty road, a malaise hung overhead. People laughed and joked around him, but it was hollow.

They filed into a building made of plywood and exposed supports, just like the place where they lived, and took their seats. Dusty floor, plywood walls. A bare ceiling that looked as if it had been put together in haste. A hush fell over them as a chaplain—not theirs, a local one from the base—began to speak.

"Good afternoon, everyone. My name is Father Matthews. I'm sorry to have to meet you under such unhappy pretense. Let's all bow our heads for a moment as we begin in silent prayer."

Half of them did, half didn't. Benjamin had flirted with the idea of being a priest when he was younger, but he didn't feel much like praying. God seemed to go one of two ways on a battlefield. Either he drew you in, or he pushed you away completely. Whoever said there were no atheists in foxholes had been talking out of their ass.

"It's always a terrible tragedy to lose someone, especially someone who's family. That's what we are here. All of us, even those of us just meeting for the first time, are members of a bond forged in fire. A bond stronger than anyone outside the rank and file can possibly know. Sergeant Griffith was—"

"I think you mean Griffin," someone shouted from the back where third platoon sat.

"Yes, of course, Sergeant Griffin. He was a good man by all accounts."

The priest lost Benjamin there. He listened as the man droned on about God's plan and the calling of warriors in a conflict to bring peace to the world. He could talk about family all he liked, but he wasn't one of them. He'd happened to be in arm's reach with a frock and a pleasant voice. He didn't know the first thing about Griff, but he stood in front of the man's brothers and spoke as though he'd been there when that shot had taken his life. It was a joke. God didn't grant that priest any special insight into what they were going through, and after almost a year at war, Benjamin didn't feel up to humoring him.

When the priest finished, Captain Fellers stood and spoke. His words were more sincere. He hadn't been a member of the company long, and the role he'd had to fill was huge, but at least he knew Griff. He'd worked with the man before. He wasn't some nobody that spoke straight from his ass when he was outside of all of it. Outside the war. Outside the grief.

The CO laughed and told stories about Griff, and many of the men in the audience joined him. When he stepped down, others stood to share their memories. Memories of smoking cigars outside their tent on BIAP. Of the time Griff, a beast of a man, had shot a Barret .50 cal from the shoulder while standing at a range. Of how he loved philosophy and spoke about it whenever somebody would listen. John Kerry had once remarked that anyone who didn't go to college bought a one-way ticket to Iraq. Griff had been proof that the line between scholar and soldier wasn't as thick as often pretended.

But soon, the memories and the laughter ran dry, and only a moment of awful silence remained. Howard stood up to tell the last story before their memorial ended.

"I'll…" He stared down at his boots for a moment. "One of the last things he did was apologize to a terp he'd yelled at earlier. Last thing on his mind was making right with someone he'd wronged… He'll always be my hero."

A lump formed in Benjamin's throat as Howard sat back down, but he said nothing. There was nothing left to say. The great tragedy of life wasn't that you died, but that you always assumed you had more time. No words of his could fix that.

Griff had remarked to his wife before his death that Spartans always came home, even if it was on their shields. He'd be going home soon. A Spartan, a Charger, a Patriot. He'd at least have that last honor of being carried back to the motherland, a warrior among sheep.

The priest shared some final, banal words before sending them on their way. Benjamin hadn't known Griff that well, but there were no tears. While Griff might be going home, they weren't. There was still work to be done for the men left behind.

For the next few days, missions wore on as normal. The battlefield remained as dangerous as ever, and the 82nd still as useless. Charlie Company continued to pull missions in Sadr City while doing half of the patrols for the other unit.

Every chance he could, Benjamin sat in front of his laptop to write, but it didn't come. He had no problem writing a letter to Tiffany every night, but whenever he sat down to work, the words vanished. He told himself it was an issue of time. He lived in a war zone, after all, working more hours every day than most people did in any given two. No need to beat himself up for a lack of creativity when he wasn't getting a wink of sleep most nights.

After the fourth night running, he decided to write to Tiffany as if he were writing in a journal. It felt important. At some point down the line, maybe years later, he would want to know what had happened. Maybe his memory wouldn't be as clear. Time would wear on him the same way it did everyone, and he wasn't going to risk forgetting about the sacrifices his buddies on his left and right had made.

He put the pen to paper and stared at the ceiling light overhead, the words staying just out of reach. Second platoon was just coming back from patrol, but other than that, the hangar remained quiet.

Nobody yelling, no loud jokes, no metal on metal as someone cleaned a rifle.

The words came slowly at first. He hated sharing the bad news with her. Things were terrible enough being on the other side of the planet, being shot at every day. But he needed to say it.

One of my teammates, Griff, died a few days ago.

It makes me feel that I am lacking to hear others speak of what a good person he was. He lived with strength and honor. He was a modern-day centurion, and I want to be the same.

We look mean, and we look tough. People avoid us on the streets here. They're scared. But these are good men. We look ten years older than we did when we started this, but I swear when I look into the faces of those around me, I see it all stripped away after this tragedy. I look into the eyes of a bunch of guys from eighteen to twenty-four, but they all look fifteen right now. Fifteen and alone.

He re-read what he'd written before crumpling it up and tossing in onto the ground below his bunk. She didn't need to know this. People bled and died here to the tune of ignorant presidents and a nation who slobbered on TV dinners. He wouldn't be their window into another world. Charlie Company didn't fight for their entertainment.

Bile rose in his throat. There was no sadness in the infantry, only anger. Anger at the MP who tried to take their guns. Anger at the people in the US, so morbidly curious, yet apathetic. Anger at the politicians who sent them here and then made themselves wealthy off the backs of soldiers.

And maybe angry with himself too. He was here, after all, taking part in a war that was moment by moment less important than it had been when they'd arrived. What the hell did they hope to accomplish here? What had Griff and Henkes died for?

He stared at the ceiling while Olson snored in the bunk below. Since he'd come to Iraq, he'd learned to sleep at will. In a place where you never knew how much you were going to get, it was a handy skill. Just close your eyes and drift off.

But right then, he didn't want to. He just looked at the chipped and half-rotted wood of the hangar ceiling above him, thinking of nothing and everything, letting the raw storm of emotion swirl inside him. It took hours to calm down, and when he did, he jumped out of the bunk and grabbed the crumpled letter. As he climbed back into bed, he unwrinkled it as best he could and placed it back inside the notebook. It might not be for Tiffany's eyes, but that didn't mean he wouldn't want to look at it again one day.

He finally went to sleep, dreaming of home and the people waiting for him.

The next day was hotter than it had been in months. He'd tried to explain it to his best friend on mid-tour leave while they browsed their old comic book store, but it was impossible to make someone who hadn't experienced it understand.

"One hundred and twenty degrees on a nice day, plus ten degrees because of the body armor. Then we get in a giant metal box without air conditioning, and we do all kinds of heavy lifting and running around. By the time we're done on a mission, you can wring out your shirt, and your sweat makes a puddle in the dirt."

Another customer eavesdropping nearby chimed in. "You're lying. You'd die in that heat." He had more chins than a Chinese phonebook and a T-shirt with more stains than original color.

Benjamin didn't miss a beat. "Maybe you would, but some of us condition for that."

Sitting in the Stryker that day, it wasn't quite so hot, but it would be soon. By the time they finished that night, he was sweaty enough. He took a shower before going to the chow hall with Simmons, Lunquist, and Howard. When they got back, the rest of the mortars stood around their bunks as if someone had shot their puppy.

His heart sank when he entered and saw it. His mind leapt straight to the worst possibility. Another man down. Another friend dead. Glancing over at Howard, it appeared he had the same thought.

Lunquist however, walked right up to Nelson. "Who shit in your cereal?"

248

Nelson took a wide, playful swing at his face and missed. "Better go ask your squad leader."

Lunquist raised an eyebrow but did as he was told. "Come on, Howard."

That Nelson wasn't angry put Benjamin's mind at ease, but not by much. "What is it?" Benjamin leaned against the side of his bunk as Simmons sat on the edge of his. Everyone frowned. Sutter shook his head and gazed at the floor. Olson pursed his lips as if he were moments from crying.

"Sergeant Miller just put the news out. They told him and Garcia today at the platoon sergeant meeting that we're being extended."

The world spun as though someone had pulled his legs out from under him. Time and again, Miller had told him to get ready for it, but deep down he'd believed it wouldn't happen. He didn't realize he'd sat down, but when he spoke again, Nelson stood over him. "Is this a joke?"

Olson jumped up and punched the wooden wall behind him so hard it cracked. That was all the answer he needed.

Chapter 24

The newscaster wore a somber expression, but Jay could see right through it. She didn't care any more than the rest of the country what was happening on the other side of the planet. The vultures on TV loved a tragedy. Every time someone died, every time a bomb went off, it validated their bullshit all the more.

"Mosul has essentially been overrun. ISIS has been overtaking city after city in Iraq, and it appears the Maliki government is powerless to stop it. It's believed that the Kurdish forces of the region will attempt to bring the fight to ISIS, but as of now, things look bleak for Iraq. It really is a tragedy."

Tragic, tragic, tragic. They repeated the word so many times it lost all meaning. But tragedy took agency from the actors in war; it didn't steal the show. Nature created tragedies, not a man with a gun. A man was only cruel. Anyone who said otherwise only excused it.

Jay turned off the TV and lay on the couch. He had to work that night and go to school the next morning, but he didn't feel up to either of them. He suspected he'd be "sick" again. Doing anything grew more difficult by the day. It all felt so…meaningless. As if sooner or later, everything in the world would end up like Iraq. All the good ever done, all the lives ever saved, would eventually end the same way.

He stared out the window into the late Tennessee summer. The green leaves swayed to the rhythm of an unfelt breeze. The sun smiled down on everything. It looked hot out, but there was something refreshing about Tennessee's warmth.

But inside, all was turmoil. Iraq had gone to shit, and Bales had been sentenced to life in prison only a few months before. The media circus around it had been maddening. Everyone wanted to chime in on things they had no clue about. Every member of the media became an expert on PTSD and war, despite none of them having any experience in either field.

Those who actually had experience were worse. Watching people he liked and respected defend a child-killing son of a bitch killed those kids all over again. Some of his old teammates spoke about understanding, reminding people repeatedly that Bales was as much a victim as them.

But he wasn't. He was a murderer. Not a killer like they'd been. He doubted those defending him would be as understanding if their wife and kids had been killed. If they'd come home one night to find Bales, bloody and drunk, standing over the corpses of their loved ones. But soldiers didn't see the foes in a war as people, so it was easy to excuse it away. They became the *other*, the enemy. *Hadj.* Killing them wasn't okay the same way it wasn't okay to kill a dog, but that didn't mean Bales should go to prison for it, at least in their minds. It made Jay sick.

On the other side of the spectrum, some of his friends had gone to Iraq to help the Kurds fight ISIS. It had been almost too outrageous to be true when Allen had told him.

"Did you hear about Wilson?" Allen asked while they sat on Jay's couch the week before, watching hockey.

Wilson had been a mortar in the neighboring Bravo Company of 2/3. A good guy. Quiet. Professional. All the things you wanted out of a soldier. Jay had forgotten about him in the years since leaving the military. But as soon as Allen asked the question, his heart sank. Every week brought news about one of their old buddies going to jail, getting killed, or losing their minds.

"What happened?"

"The crazy bastard snuck back into Iraq."

The news hit Jay like a bucket of cold water. "What?"

"Yeah, crazy, right?"

It was crazy, and a week later, it hadn't grown any less so. Jay stared at that picturesque scene outside his window, and deep down, he felt he should be there too. It was absurd, of course. He'd been a soldier before, backed by the most powerful military on the planet. Now he worked at a hospital. Just a guy passing clamps to overpaid surgeons. But he and his friends had a hand in the events even now rampaging across Iraq, and he couldn't help but feel he had some responsibility to help fix them.

Half a dozen times that week he considered finding Wilson's family to discover how he'd managed to sneak back in. He certainly hadn't gone in through BIAP. It would be easy enough to sneak into a neighboring country, but how the hell could you find the Kurdish people from there? What would stop you from wandering into ISIS instead of those you were trying to help?

Not three days after Allen had told him about it, Wilson appeared in an article on the internet talking about veterans helping the Kurdish. He spoke of their responsibility. Of how it was the right thing to do. Every word he'd written was a knife in Jay's guts. By the end, he could only stare at the picture that had come with it. Wilson, still short and just a little heavier. He wore dirty camouflage, and the dusty, broken-down buildings behind him were exactly how Jay remembered Iraq. Kurdish soldiers surrounded him. Dark brown skin, dark hair, and an expression somewhere between anger and resentment on all their faces, as if they were upset at the audacity of being reported on instead of supported.

Wilson had the courage to go back and take responsibility, while Jay lamented the fate of the country he helped ravage. Again and again, he tried to tell himself they'd been there to aid the people, that they'd done good. But invariably he turned on the news and saw the results of their particular brand of help. Did Wilson have the right of it? Could you fix the problems created by violence with more violence?

Jay pulled his cellphone out of his pocket as those thoughts swirled round and round in his mind. "Lucy? It's Jay. I'm not feeling really great again today."

It went on like that for weeks. The lack of sleep. The rush of negative thoughts. Every time he thought he'd escaped it, they came back stronger than before, waiting in the wings to swoop in and destroy any joy or happiness he found. The news continued to report on the unfortunate situation in Iraq; the tragedy, as they called it. Jay couldn't look away. He wanted to, but something inside made him stare. If he wouldn't go over there to help Wilson, the least he could do was keep a watch from home.

And so he paced back and forth at night, wondering, wondering, wondering. Muttering under his breath about how things might have been different if they'd done things just so. If that man hadn't said that. If this president hadn't decided this. Stumbling over the chain of events that led further back than he could fathom. To days when men killed one another with swords. Before that. To times when rocks were weapons and reasons were nothing more than impassioned grunts and a hard stare. Always wondering who to blame, and if it should be him. A maybe-not-so-innocent man who'd made a decision as a teenager that would affect him forever. A not-so-innocent man who'd kill other not-so-innocent-men while the presidents and terrorists actually responsible slept peacefully in their beds, blissfully self-assured that every move they'd made was the right one.

Allen noticed the stress the next time they hung out. "Have you considered going to the VA? Maybe one of the docs there can help you get your head on straight."

Get your head on straight. As if he could walk into a clinic and someone could make it all better. Jay bit back the response that threatened to bubble out in the most condescending tone possible. *Thanks for the advice, buddy. Cram it up your ass.*

Maybe Allen had a point. He couldn't keep on like this. His grades slipped, gradually at first, then much faster. His attention span had vanished, and everyone at work knew something was wrong. They were polite enough to whisper it behind closed doors, but hospitals were small, and rumors got around.

He didn't want the VA's money. He didn't want anything from the government anymore. All of it felt like bribery. Those military paychecks, that overseas tax-free money he'd made as a contractor, the free food, the housing, and now the VA disability that so many of them took afterward. Some said they'd earned it, some said they didn't. No matter which way you leaned, it was all blood money. Burn down a country and take the prize. Never mind the millions who suffered. Pay no attention to the other side of the planet where people hid in their homes, cowering behind a single AK-47 and hoping no more misery rained down on them.

The semester ended, and Jay sat inside the office of the professor who ran the honor's program. The office was like the man—tall, lean, dignified. Diplomas hung on every wall to reflect his many accomplishments.

But all hints of the kind, mirthful man who'd extended the invitation to the program only a couple years before were gone, replaced by the one who looked at Jay over steepled fingers. "Your grades are slipping. We're going to put you on probation. If you don't get them back up by the end of next semester, we're going to have to remove you from the program."

He considered explaining himself but decided against it. The rules didn't care about him or his problems, and telling his not-so-sob story wasn't going to keep him in the program. He thanked the professor and left.

He didn't care about leaving the program. Picturing the future became harder every day. Grad school? A good job? All those thoughts remained distant. He couldn't care about any of that when his old friends were sneaking across borders to fight the enemy they'd made together.

But that was the wakeup call, that he didn't care. As he climbed into his car to go home and do nothing once again, he knew it wasn't right. He drifted by those young, smiling faces on the campus. The quad with its red brick buildings that had once screamed promises of what the future might hold only looked like red brick now. The magical late summer Tennessee had grown mundane, no longer ripe with possibilities about what might be right around the next corner, now it was only hot.

When he returned home to his mostly bare apartment, he looked up the VA number and called. Blood money it may be, but maybe they offered more too. He wasn't so naïve to think a pill could solve his problems, but at the very least, counseling might offer him a little peace of mind. Anything was better than feeling miserable all the time.

He called, and they informed him it would take weeks, maybe months, before he could see anyone about his problems. Instead, they asked him if he thought about hurting himself. If he needed immediate medical aid. They spoke to him as if he were already holding a gun in his hand.

Regardless, he made his appointments. He swallowed his pride, staring out the window at the bright, sunny day that felt so very different from his mood. Looking out that window, he came to a decision.

Instead of picking classes for his next semester, he wrote an e-mail to Professor O'Leary thanking him for experiences of being part of the honor's program.

I really appreciate the opportunity, but I think I need some time off. I'm dealing with a few problems, and I don't want it to continue affecting my GPA.

The written word from the warden that it was time to execute the prisoner. He'd been in school long enough to know most people who dropped out in their senior year never bothered to come back. He just didn't care.

As he hit the send button, the last vestiges of his dignity finally left. He sat there on his ratty couch, leaned forward, and cried.

Even in secret with nobody watching, it felt shameful, but he couldn't help it. Once the heat behind his eyes had coalesced into those first tears, the rest followed. Nothing he'd ever done mattered, and now he'd given up.

The tears didn't last long. Shame drove them away. The wondering what his friends would say if they could see him now. Imagining the look of disgust on his past self's face, his soldier self.

Cars drove by outside as he shut the blinds, closing himself in to the comforting darkness. He felt like a bag of broken glass, poked full of holes and filled with sharp edges. He did nothing but stare at the wall for hours, until it was finally time to go to work. But he couldn't. Or he wouldn't. He wasn't sure which.

He called in once more. "Hey, Lucy. It's Jay. I'm not feeling well again today."

His boss was one of the nicer people he knew at the hospital. Older but still full of youthful energy, she never had anything but a smile on her face. Today, however, was a day for upsetting happy people. "Jay, you can't keep doing this. If you call out today, I'm going to have to fire you."

The air vanished from the room. His chest tightened, and all the sound in the world stopped. "What?"

"We've had to have others pick up overtime to cover you the last few weeks, and it can't keep going on. Unless you can go to a doctor and bring in a note saying something is actually wrong, I can't let you off today."

There had been signs, of course, and she had spoken to him about it on a few occasions over the last month. But he recalled it through the same fog as all his classes over the last year. Had it really been so long since he gave a damn?

The words that came out of his mouth next surprised him. "Do what you have to do." What was one more bridge burned? What did any of it matter anymore?

"Are you serious?"

Instead of responding, he hung up. Being the responsible adult the world expected him to be wouldn't change anything. Everything would go on as it always had. Nothing he or anyone else did mattered.

He paced back and forth as the beginning of a panic attack took root in his guts and worked its way through his body. They were becoming a weekly occurrence. More often sometimes, as if even his body rebelled as he ruined his life. His fingers tingled. His heart rate doubled. More than doubled as his breath escaped him. His vision narrowed, focusing in on what was right in front of him without seeing it.

Just a phase. Just a passing thing. Nothing to worry about.

He'd picked up that trick last spring. If he spoke to himself, it would go away. They were transitory. They were all in his head, nothing more.

But the panic attack continued. Crumbling the walls and raining misery on him. He started speaking aloud to himself and the four walls. "You're just having a panic attack. It's been a bad day. It'll pass. No big deal."

But it wouldn't pass. He'd quit school and his job in one day. He had savings, but they wouldn't last forever. Without it, he'd end up on the streets. Was that how it happened? Was every homeless veteran just a person who'd had a bad day and threw it all away?

He kept pacing until the light under the drapes had vanished, but even once night had fallen, there was no relief.

People always said something stupid in movies when they held a gun.

"It's heavier than it looks."

"It's so big."

Not him. Maybe it had always been there, or maybe it had come from years of training, but a gun felt right in his hand. His father, and his father, and his father had all been soldiers. Killers. Maybe it was natural.

Those thoughts and a thousand others raced through his head as he sat on his bed, staring at the gun in his lap. A Christmas present from his mother only a handful of years ago. It had sat in his closet under a pile of clothes and old boxes for most of that time.

Allen had asked him about that once while they were out drinking. "Not going to do you much good under the bed, is it?"

And Jay had laughed. "I'd rather kill someone with my hands."

But it was hard to kill yourself with bare hands, and he was past the point of waiting for it to happen by itself.

The thought shocked him out of the deep thoughts. Had he really come to that point? He'd thought about it often enough since the panic attacks started. Since the world stopped looking like an oyster and more like a steaming pile of shit. But to be sitting here, gun in lap?

In hand. He picked it up and dropped the magazine from it for the fifth time in as many minutes. A full magazine. He should have taken at least half the bullets out years ago. Someone had once told him that too much pressure over too long a time could warp the spring in the magazine. Someone else had told him that was only an issue with the old, shitty magazine the army used. He had no idea which one was right.

His thoughts kept rambling. He blew out a sigh and stood up, pacing back and forth in his small bedroom. Clothes covered the floor. Trash. It hadn't started to smell, but it was just a matter of time. He hadn't left the house in days. Or had it been weeks? He'd lost his job, quit school, and suddenly time didn't seem as important as it had before all of that. As if the only thing keeping him glued in the here and now was his title or the one he would have with the education. Without that identifier, he came loose. A shingle on the roof of life, unstuck in the storm.

Or maybe those weren't the identifiers whose loss he felt so keenly. Maybe it was a different one from when he'd barely been a man. When he'd travelled across the world with his family, not the one he was born to, but the one he chose. Maybe he'd been unstuck since then. Lacking belonging in the most sacred secret club on the planet.

PTSD is the realization you'll never be as cool as you were when you killed people.

He'd laughed at it then, but he didn't now. Instead, he stopped pacing and stared out the window. The sun had begun to set, painting the world in crimsons and golds. At this time of year, the mountains outside of his apartment were pregnant with life. The world outside filled itself with possibility.

But not in here.

Not anymore.

He cocked the gun and stared down at it again. He's spent years cutting himself off from every stable shore. Teach them how to kill. How to hate. And where did it end? Big, bold letters spelling out "The End" in the clearest way possible. Some upstairs neighbor or another would hear the shot and come look. Others would gawk from the stairwell outside, catching a glimpse of the body covered in a black tarp as it was wheeled out to the ambulance to be dumped in some local meat locker. A month later, they'd finally find his family, and weeping people who only called on Christmas and birthdays would file through, picking up his things and dividing them amongst themselves. In ten years, nobody would remember the sound of his voice.

Well, not nobody.

Donner would. Brenner. Nelson. Allen. Duarte and Simmons. Ski. Masters. Sutter. All of them. They would.

And just like that, it *was* heavy. He looked down at it again, and the fear that had been nowhere to be found filled him with horror so pungent he dropped it. It hit the floor with a thud strong enough to shake the picture frame on the wall next to him.

All of his friends would remember. They would remember and blame themselves. They would drink to his memory in some distant future and wonder why he hadn't been strong enough to call. They'd shake their heads and ask one another why he hadn't just reached out to spare them this grief instead of laying it on their doors.

A voice—only imagined, yet strong enough that he couldn't ignore it—spoke to him. "What are you doin', man?"

No, that wasn't right. The words whispered in his head were the ones Kennedy had parted with the last time they'd seen one another. "I want to hear about all the best-sellers you write when you get back home."

This wasn't right.

He picked up the gun and dropped the magazine again before pulling the slider back to release the bullet inside. He gathered them all up, even the ones buried in his closet, and threw them in the trash. When he'd done that, he grabbed the bag, took it out to the dumpster, and threw it away, gun and all.

A weight lifted off his shoulders. Close call. *Near miss*, as they would have said in Iraq. No shots fired, but damn if someone didn't almost die.

He laughed at the dumb joke. The teenagers on the porch nearest to him stopped to stare at the strange man chuckling at a dumpster. But he didn't stop, just let it slowly peter itself out until he was only smiling. Then he stopped that too, and the quiet monsters lurking in his head settled in once more.

He pulled out his phone as he walked back to his apartment, calling Allen. "You busy? I could use some company."

Chapter 25

The explosion that crumbled the building rocked the Stryker from a block away. Bombs often felt like an earthquake. Dust shook loose from the panels around Benjamin as his heart skipped a beat.

"What the fuck was that?"

Before Nelson could answer, the company radio came to life. "The house that first squad was in just blew up. They're…they're coming out."

Benjamin couldn't see a damn thing from his hatch, and he wasn't allowed to open it unless he had to. The risk of a driver being killed and a vehicle being stuck somewhere was too great. But through the periscopes, he could only see the cracked, gray wall directly in front of him.

"Sergeant Nelson, can you see anything out there?"

It took him a moment to respond. "They're carrying someone. It looks like Jaiden, but he's moving."

Benjamin's chest tightened as sweat ran down his face endlessly.

Now that spring had sprung, the temperature skyrocketed. Just as it heated up hotter than the devil's dick, they'd stopped working with the 82nd and gone back to BIAP. It looked for a while as if it was good news. They'd been covering down on two units, after all. Everything in Sadr City had reeked of shit more than most other places in the city. The people in that garbage-filled cesspool hated them, they'd pulled more missions, and they'd lived farther away from everything.

Somehow, the army had worked its magic and made everything worse. The mission tempo increased. Attacks doubled as the troop surge continued. Thousands upon thousands more soldiers arrived in country every week. The chow hall lines grew outrageous, the battlespace was always full, and over all of it hung the death sentence of the extension. They'd been so close to going home, only to have it pulled out from under their feet.

And now one more of them might not make it.

"He's moving around," Nelson said. "It looks like—"

The first sergeant barked orders over the radio. "Get him into the MEV, jokers." Orders and exfil[64] plans went into effect. Everyone in the company, even the leadership, had long passed the point of putting it all on the line. Those on their first deployment were experienced soldiers by now, and those who had been through

---

[64] Exfiltration

many were tired. The infantry had to get its job done, but your buddies always came first. The mission had almost been over anyway. No weapons caches here. Just more angry, dirty people, staring at the Americans as if looks could kill.

By the time they returned to BIAP, 1SG had confirmed that Jaiden would live, but he was wounded. The first thing he'd wondered after the initial shock of the extension had worn off was whether they'd lose people during it. To receive that news so soon after Griff had died was cruel in the way only coincidence could be. Jaiden still lived, and they weren't technically biting into the extension yet, but that thought still plagued him.

Jaiden's platoon took him to the hospital on base while the rest of the company returned to their quarters. No more tents or hangars. They lived in CHUs a lot like the ones they'd been in while they stayed in Mosul, and right up the road from where they'd been before. Going in circles, ever the military's philosophy.

He climbed out of the Stryker as the rest of the company dismounted. Even on the best of days, they looked tired. When they'd first arrived in country, it seemed nothing like a real war. It still didn't in many ways, but they were real soldiers now. Hollow-eyed and scrawny from too much work and sweat. In all the pictures he'd seen, World War II soldiers had the same eyes, the same demeanor. If he could travel through time and see those people face to face, he didn't doubt they would recognize one another.

Al Barrera

"Do your PMCS[65] and get back to my CHU in half an hour for a meeting." Nelson took off his kit and put it away in the truck.

"Roger, Sergeant."

Vehicles required endless maintenance. They rode them hard, and precision instruments broke down under the best conditions. Fluids, tires, and machinery needed to be checked constantly. Comms and weapons had to be maintained. At the best of times, it was a hassle. Here in the desert, things broke twice as often.

But it had grown mechanical. Benjamin instead let his mind drift to Jaiden. They hadn't said exactly what happened to him. Benjamin recalled all the times living back in the barracks he'd seen the man. Watching him practice backflips in the day room on base. Other members of his squad laughing and calling him cereal bowl chest because of the way his sternum bowed in slightly. Jaiden laughing it off. He was handsome, and he knew it. He aspired to be an actor when he got out.

Could he be an actor now? Would his wounds end that for him? Would he forever lament that moment when his future had changed? Benjamin could wonder the same thing about all of them on any given day, but that didn't make the question any less pressing.

His wrench slipped off the bolt on the c-diff below the truck. His knuckles knocked against the metal, scraping the skin off and leaving a bloody gash. "Fuck!" His temper flew away from him, and

---

[65] Preventive Maintenance Checks and Services

everything became white hot rage. He pulled the wrench off the bolt and hammered it against the metal of the Stryker above him.

Ting! Ting! Ting! Ting!

Each strike left a tiny dent and scraped paint away. Miller always told him if he broke anything on the truck he would have to pay for it, but that was as bullshit as everything else around here. They couldn't punish him more than he was punished by being here.

Ting! Ting! Ting! Ting!

Nothing good happened here. Every day the threat of death, either his or one of his friends, hung over them like a shroud. It was never a question of if, only of when. He'd been a naive fool to think they would get through this year—these fifteen months—without any of them dying or wounded.

He stopped pounding and just lay there, sweating like a pig, staring up at the belly of the beast that kept them alive when they should have been dead a hundred times over. Had he really thought that they'd all make it? He had believed there would be no going home for him when he'd stepped off that plane a year ago. Could you reconcile those two ideas?

He dropped the wrench into the sand and gravel under him, and it landed with a clatter. He didn't know anymore. He didn't know anything. It was like walking in the dark. He could only put one foot in front of the other and hope he didn't fall into a hole.

Olson's head blocked out the light from between the tires as he bent over to look at Benjamin. "You okay down there?"

Benjamin glanced at him. Their spot lay at the edge of the parking lot. Beyond the tires, he could see all the other Charlie Company Strykers down the long line to the edge of sight. All the other drivers and TCs did the same work, moving around, going about their business. Busy little bees making their honey.

"I'm fine." He picked up his wrench and set to work tightening the bolt again. "Just a little pissed off."

Olson stood and mumbled, but Benjamin caught his words. "Me too, buddy. Me too."

It took a full day to receive the news about Jaiden. A patrol gave way to a lot of sitting around the CHU and looking at pictures of Tiffany on Myspace. They'd installed internet connections in their new homes, and Miller had droned on and on about how lucky they were to have something like this. How they couldn't possibly understand how good they had it. It reeked of the *I walked fifteen miles uphill in the snow to get turned down for a job!* bullshit so many loved to spout. Time moved on, and things changed. In a hundred years, wars would be fought from the other side of the planet by remote control. Some haggard NCO would tell his privates and specialists they didn't know how good they had it being able to go home to their wives and husbands every night, and it would be just as bullshit then.

They didn't suffer as much as people had in the past, but they also ran more missions than any military had in previous wars.

Brenner kept the count for fourth platoon. There were hundreds of missions for them alone. When you factored in the company ops, the count went through the roof. They ran more than a mission a day on average. And for what?

No sooner had that thought occurred to him than Nelson burst into the room he shared with Sutter, Ski, and Simmons. "Squad meeting, mortars. Get your lazy asses over to my CHU." He vanished back outside without closing the door.

"You really should housetrain him," Ski said from over his Gameboy.

Benjamin, Simmons, and Sutter trudged through the gravel to the other side of the lot where Nelson shared a room with Olson. The same T-barriers closed them in as everywhere else. Unlike back in Mosul, there was a lot less mirth. People still smoked and joked outside like always, but it was more subdued. Soon they would be cutting past their allotted year and into their extra three months. Rumors flew that it would be extended again to six months. The real pessimists among them said they would get the horror story to end all horror stories and end up here for two years.

A month ago, Benjamin would have laughed it off as impossible. Not anymore.

They entered Nelson's CHU to find the rest of the mortars already there. Everyone wore their PT uniforms, showing off the farmer tans the lighter of them had acquired. Masters had once remarked that they looked like a bunch of half-baked bread rolls, and Benjamin couldn't disagree.

"Get in here and close the door." Miller never said anything without a shitty tone when he could avoid it.

They obeyed and sat on the floor, reminding Benjamin of story time as a child. Kids huddled around the feet of the adults, listening with rapt attention to every word. Benjamin hated it, but Miller loved to reinforce that idea.

"Jaiden had his back broken. He's going to live, but they need to evac him to Germany for surgery. Deployment's over for him."

Benjamin blew hard out of his nose. Relief flooded him to hear one of theirs would be okay and going home early; it was just one bummer of a way to get there. "Did they say anything else about him?"

Miller turned his Coke-bottle gaze on Benjamin. "If they'd said anything else, I would have put it out, wouldn't I?" He stared at Benjamin as if waiting for a response, but there was no good one. Miller was just as miserable as the rest of them, only he opted to take it out on his subordinates.

That same flash of anger that had overcome him as he beat on the Stryker the day before threatened once more, but he swallowed it. He just had to keep putting one foot in front of the other. One day at a time. A minute at a time. A second. They had to do it if they were going to make it through. It would be over soon, and everything would be okay.

Or not. They would stay as long as the army needed them. Well past the point where anybody cared about the war, they would stay.

Miller finished the meeting without putting anything else of substance out, and Benjamin managed to bite his tongue. Back talk only validated Miller. Instead, he grabbed Masters and Donner and made his way to the chow hall before their company mission that night.

TCNs maintained the base well enough, but dirt, gravel, and sand made the going slow. "Can you believe that guy?" Benjamin asked as they trudged across the long, open stretch of sand that separated the chow hall from the tents.

Masters shrugged, his weapon dancing on the end of its sling. Sweat poured off his brow from their short walk. Summer would be miserable. "Guy's a dick. Fuck 'em."

"He's not your problem anyway, short-timer," Donner added.

That much was true. Benjamin would be getting out only three short months after they returned home. That still put him in stop-loss[66] territory, but not as bad as some of the guys. Allen Rosenberg was supposed to have gotten out a month before they left on their now fifteen-month deployment, and they weren't allowed to ETS[67] until three months post.

Supposedly, it was a measure to see if you would go crazy after a deployment. Just giving soldiers some cool off time before releasing them into the wild once more. Making sure they didn't beat

---

[66] When a soldier is kept past their contract date due to a deployment or injury.
[67] Expiration of Term of Service

their kids or drink a fifth and drive off a bridge. Three months seemed too short a time to tell something like that, but the military loved its redundancy. After all, Vietnam guys still lost their minds out of nowhere.

Would that be him and his friends in forty or fifty years? Everything going fine, then *bam!* Suddenly you didn't know who you were or where you were going. You ended up on the street or in some rest stop bathroom with a needle in your arm to make the memories stop.

He didn't want to dwell on it. "You should come with me. We'll go start that porno company you always wanted, Masters."

Masters collected porn, and he swore up and down he would direct it one day. "Hell no, dude. Gotta get that military pension so I have something to fund my retirement hobby."

Donner laughed, and Benjamin zeroed in on him. "We can have Emilia star in your first film, Masters. We can call it *Army Wives: From Green to Black*." Benjamin waved his hands as if he were framing a marquee.

Masters laughed, but Donner was untouchable. Nothing ever got his goat. "I could only ever marry a woman who's had all her holes plugged at once."

They all laughed together, and the tension from the meeting slowly drained away. That constant underlying amount couldn't be helped. Much like the national debt, you could never erase all of it. The trick was to minimize adding to the pile.

The next day grew hotter, and fourth platoon went on a longer patrol. They took small-arms fire on some back street in a terrible Baghdad neighborhood, not that there were many great ones.

"We could just drive through all their market stands next time we come down here," Benjamin suggested. The stands were vacant now. As much as the US liked to talk about winning hearts and minds, these people had to spend all day around terrorists. When the bad guys said, "Get lost," you got lost.

"I might want some of that foot bread tomorrow. Pass," Nelson replied. People stomped on the dough and then put it in a wood fire oven. Benjamin had his doubts as to whether or not they actually stomped it, and he'd never seen it made, but he couldn't deny the taste.

There was no bread that day or the next. The following week, the stalls remained closed. Covered overhangs on the side of dirty buildings on narrow, garbage filled streets. More and more, the missions happened at night, though a fair share still occurred during the day. They worked with civilian contractors to put up barriers around a particular neighborhood to restrict access and net the terrorists in. Nobody stated the obvious. That if it took weeks to put up these barriers, all the bad guys would just take their toys and go to a different area before it was finished. No amount of all day security between Charlie and Bravo Companies could stop it forever.

Pointing it out wouldn't change anything. Why bother? If it wasn't this pointless mission, it would be another. Some officer even now sat somewhere beating his dick in a mirror while thinking of what a brilliant fucking idea he'd had.

Night after night, the company trucked out to pull security. Boring work occasionally intersected with bouts of gunfire or minor IEDs. Par for the course at this point in his career, so much so that he wondered what it would be like to finally be out. He had plenty of time to think, after all. Would he miss it? Would his life ever be this interesting again?

He sat in the driver's hatch with nothing but the glow of the DVE to see by. It was hard to picture a job worse than this. Miller went on and on about how much more difficult the civilian world was, but Benjamin doubted it. Having to get yourself up and answer to a boss that treated you like shit wasn't so far removed from his current situation.

The next day, the company loaded up to clear out a bunch of houses in some shit-stinking neighborhood downtown. This time, his Stryker faced a street instead of a wall. Not many people wandered the roads. The three- and four-story buildings left deep shadows on the roads. Children played in the garbage-strewn alleys. The sweltering heat brought out the smell worse than ever. When the Americans had come to Iraq, much of the infrastructure had been destroyed, including sewage. Somehow, they hadn't yet fixed it in many places. Some conquering heroes they were.

Little kids swung sticks at one another, playing as if they were swords. He'd done the same when he was their age. He wondered what they thought of the invasion. Some of them weren't old enough to consider it much more than a fact of life. The Americans had been there forever and, as far as they knew, always would be. It fit with what other Iraqis thought. When they'd questioned a local about the army for fun, he'd been all sorts of wrong. He thought the military could read their minds, see through walls, and control their thoughts. Benjamin and the rest of the guys in Quinn's squad had laughed about it. Even Brenner had gotten in on it.

"That's right," he'd said. "So you'd better be on your best behavior."

It had been funny at the time, but eleven months into the deployment, it wasn't. These people had no idea what was going on around them. They played either victims or terrorists come to chase the "Western devils" from their land. As the children played in the shadows of the street, he couldn't help but feel they were the biggest victims of all.

Boom!

An explosion somewhere up the street shook the Stryker.

Company radio squawked to life. "Charger elements, charger six, status, over."

Ayers, one of the first platoon drivers, came over the radio. "We got hit. Clint's hurt." His voice shook, reducing him from the twenty-something man he was to a boy.

Captain Fellers stayed calm, the mark of a good officer. "All right. We're on our way."

They exchanged information back and forth. An IED had gone through the bottom of the Stryker and hit the VC[68]. Everything else remained hushed over the radio.

"Fuck," Nelson said.

They were all thinking it. For two of their own to be hurt so close together, for it to happen so close to Griff's death and the extension...

"Man," Masters muttered into his mic, "Fuck this place."

A few moments later, they'd loaded Clint up in the MEV. With a vehicle down, the mission was a bust. It took a platoon to bring a broken vehicle back to base. One had to tow, and the rest had to escort. Down a whole platoon, they couldn't complete large-scale company operations. Sometimes they could muddle through, but this wasn't one of them. They exfiltrated back to BIAP.

Silence marked the ride home. Simmons and Masters tried to start a conversation a few times, but the quiet drowned them out.

They returned to their CHUs and went about their day as if everything was normal. As if one of their friends wasn't hurt, again. They continued to pretend the world turned like normal in this shithole warzone.

---

[68] Vehicle Commander

On his way to the shower, Benjamin ran into Doc Lars. He'd been on the Stryker with Ayers and Clint. He must have been the one to treat their injured friend.

Lars sat in the shade of a T-barrier just outside the company area. Blood still stained the cuff of his pants. His rifle leaned on the barrier next to him. He wasn't a smoker, and there was no drinking here, but he looked like a man that needed both. He searched the open area next to their quarters. Nothing but an empty road, a few palm trees, and a lot of sand.

Lars clearly didn't want to be bothered. Benjamin wouldn't ask about Clint, but he couldn't just let one of his brothers sit alone after something like that.

"How you holdin' up?"

Lars pulled his gaze from the horizon to Benjamin. "I'm fine."

The ever-present lie of the soldier. But what else could they be? Being anything but fine didn't make it all easier to bear.

"Clint will be all right, man. He had a good medic."

Lars returned his eyes to the road. "He's gonna lose his lower leg. The IED went off right under him. I got the morphine in, and I got a tourniquet on there, but there's no way he doesn't lose some of his leg."

Benjamin stood dumbfounded. He hadn't asked. He hadn't wanted to know. Having no clue what else to do, he sat next to Lars and watched the sun set with him. Neither spoke another word.

Chapter 26

His old brothers posted pictures of skulls and crossbones, of men with guns accompanied by trite nonsense. "Everyone wants to be a lion, until it comes time to do lion shit." It made him angry just to look at them. What were they trying to say? That the destruction they'd wrought was justified? That it had all meant something? The temerity of it turned his stomach.

Or were they just looking for meaning like he was? Were they uncomfortable, disconnected from everything around them, trying to express themselves in waves and grunts? Trying to find some comfort in a world that was as empty as the pictures themselves.

Soldiers were a breed all their own. Strong. Idealistic. Innovative. Young men and women looking for a cause to champion. Defenders of democracy and the right to peace and happiness. But those ideas couldn't be forced on others, and as so often happened along the righteous road, something went wrong. Any idea taken to its extreme was extremism, no matter how noble it might have been on the outset. The fighting in Iraq had long since ceased being about right or wrong, if it ever had been. It had become about *who* was right and wrong. About capitalism, and swinging dicks like swords. Soldiers might be the best and brightest, but even they could be used for ill purposes. No amount of roaring after the fact would fix it.

He closed his laptop and stared at the wall, letting the December chill that had seeped into the apartment sink in. The VA had come through for him. He'd spent a close few months wondering how he was going to keep the lights on. Every week became a question of food or bills. His dad had made a comment to him once that the working-class world was all about that. He hadn't believed it then, and he wasn't really working now.

He hated that thought. In his younger days, he'd believed the US welfare system was unfair. How could some sit at home and do nothing while others worked for a living?

But he'd tried. A two-week foray into working at a restaurant when the bills had become too much to bear turned into a screaming match with a manager and storming out in the middle of a shift. Everything he did elicited a panic attack that he couldn't escape.

He'd certainly gone more liberal over the last few years. "Gone native," as some would say. His thoughts on what the country should do for those in need had softened, and even though he thought they should limit government assistance, he didn't think it was the great evil he'd been raised to believe.

But the government forked over cash for someone else, not him. He'd been a warrior, and he was going to be a doctor. When that fell through, he'd still had lofty aspirations. And now what did he have? An empty apartment, a VA appointment every few weeks, and some money to keep him from ending up homeless. He hated it. He hated the very idea of needing it.

But here he was.

His friends tried to tell him he was thinking about it wrong. Some had seen it as a reason for him to celebrate. After all, he didn't have to worry about it anymore. His money situation was taken care of, or so they had said.

Allen had jumped on that bandwagon when Jay had called and told him. "At least you won't end up sleeping on my couch."

But it wasn't his money. It was someone else's. Charity. He'd gone from the example of what every red-blooded American should be to the poster child for the ravages of war on the human mind. He'd served as priest in the civic religion, and now he was just street trash.

It didn't help that the news always hammered home issues about the war, even after its completion. Some said it was a nightmare; others said it was necessary. Tony Blair had gotten on TV and denied that he and his soldiers had anything to do with the creation of the current ISIS, but a blind man could see they had. The ISIS leadership said as much. That without the US in Iraq, there would be no ISIS. That they'd all met in American prisons and plotted where they were going from there.

Those places had always been breeding grounds for terrorism. The open secret that the US was torturing its prisoners had come out. Even Jay had been shocked at the details of it. People hung from ceilings for days on end. Left to freeze to death. Made to starve.

Cheney jumped on TV to try and defend it. "My definition of torture is watching those planes crash into the World Trade Center on Nine-Eleven."

But it was bullshit. He defended his involvement in one of the most monstrous wastes of US resources to ever occur. His friends at Haliburton had made themselves richer than they could have ever imagined, and US oil barons had come out like kings. The US had dragged its name through the mud by defying the UN and torturing people, and again by pulling out too early and leaving Iraq in the hands of men worse than the ones they'd gone there to fight. Countless lives had been lost.

And all for what? The idea that the US had to prove it was the best in the world by beating down the doors of anyone who denied it? For the flag and people? The flag didn't care. No matter what it represented, it was just a piece of cloth. You could light in on fire, throw it in a barrel, and roll it down a hill. It didn't change anything.

That left only the people, and most of them hadn't supported the war by the time Jay and his brothers had gone there. They didn't support torture. Most of them didn't seem to give a rat's ass what happened to Iraq, Afghanistan, or any of the people in it.

The people of the United States seemed to care only for themselves, unable to see their place in the larger picture. Unable to see the bread crumbs of history that had placed globalism at their feet, whether they accepted it or not. All the men and woman back home only looked at the ground in front of their feet, occasionally peeking up at the bright lights on the horizon to be blinded by dreams shaped like dollars; the tall signs that said, *I am Wal-Mart. Look upon me ye mighty and despair.* It sickened him.

So what had they fought for? What had his brothers been wounded or died for? Why was he now one of the grasping, hungry people living on the government dime?

He left the apartment to go for a walk. His doctor at the VA had told him it would help.

"When we stay inside all day, we tend to stew in our own bad thoughts. I'm not going to tell you to force yourself to go out with people, but just going for a short walk in the sunshine can do wonders."

He was seldom wrong. Every piece of advice he'd given Jay had been helpful; but it wasn't always easy to care. That was the insidious thing about PTSD and anxiety. They took your will. They made you cease to give a damn about anything going on around you.

Or maybe you cared more than most, and it made you feel you had to pretend you didn't to survive another day. He still hadn't figured that one out.

The sunshine warmed his face, and the fresh air tasted wonderful. People rode by on bikes or jogged past him going the other way. Even in winter, Chattanooga hummed with energy. Nothing as paltry as a chill could stop a city on the go. He needed that sometimes. His life had come to such a grinding halt over the last year that it was hard to reconcile it with the hard-charging future doctor he'd once been.

Only a few years past, he'd come back from a job overseas and taken five classes every semester while working a full time job. Only a few years before that, he'd been a warrior. It was hard to picture now. He'd been little more than a child then. Many of them had been when they went to Iraq.

And so what if some of them had been boys when they went to war? What was a man? The people they became sprouted from loud, obnoxious children playing soldier. But by the time it was done, they weren't playing anymore. Maybe they'd changed when they came back, becoming something other than the killers they'd been. Maybe they'd moved forward to make millions as businessmen. Maybe some went back to who they were before. Just working-class people quietly fading into the machinery of society. Maybe some broke down as he had. Regardless of who they'd become, a piece of them never left Iraq. Those shards infected that land as surely as the land infected them.

What they were had been real, though, at least for a time. Real-life supermen with nobody to rescue and nothing to stand for. A part of them would always be that. Maybe that was why it was so hard to look in the mirror now.

PTSD wasn't flashes of combat, at least not for him. There were memories of the days spent in the back of a Stryker, of course. The months, the years. Of blood stains on the ground and friends there one moment and gone the next. Images of little Iraqi children playing in the street, and a recollection of heat so deep and hard he was certain it would kill one of them eventually. But that wasn't PTSD for him. For him, it was a voice. Just the ghost of a whisper in the quiet hours of the morning that reminded him of the big, eternal truth. One day, all those pretty things he had would be gone. He'd be old and rotten, falling apart and dying just like everyone else. It would mean nothing. It would come to nothing. The generation to follow would forget, or fail to learn, just as he had. A generation lesser than the one before it, poorer for wisdom eternally, just as his father had said, and his had said, and his had said. And long after he'd died, that voice would still be there, laughing as it claimed the next soldier in the next war for freedoms that never appeared.

Or maybe not. Maybe their children would be different. Maybe the next generation would be the one to realize the folly of slaughter before instead of after. He didn't know, and he pushed it out of his head as best he could, opting to enjoy the beautiful day while he had it.

He returned home and set to work writing as he did most other days. He could hardly call it a job, but it was something to do. He sat in front of his computer and typed away at the novels he'd always said he would write growing up. Miller had once told him that only people who'd already accomplished something became writers. Writing was for those at the end of a career, not a twenty-something who only thought he had important things to say and stories to tell. But if he didn't have this, he didn't have anything anymore.

It wasn't much, but it was something.

Sometimes nameless fears chased him down streets he thought he knew. Other dreams found him among his old brothers again. Those were the strangest. He remained the man he'd grown into over the last eight years, and they were always the same ass-kicking, no-nonsense soldiers they'd always been. Picaresque killers with a smile and a mission. He didn't fit in with them anymore in those dreams, but he tried anyway. He went through the motions and spoke like one of them, but everyone knew he didn't belong there.

His phone vibrated on the bed stand when he cracked his eyes open. Daylight spilled into the bedroom. He kept the curtains drawn, but a small amount always found its way in.

The only people who called him anymore were the VA staff, usually to remind him of an appointment. He grabbed the phone and hit the button that lit up the screen.

It was a message from Sutter.

Tell everyone I'm sorry I wasn't strong enough.

He cocked his head to the side like a stupid dog, not understanding in his half-awake state. He read it again.

This time his heart leapt into his throat.

Sutter had gotten out the same time as Jay and gone back to Texas. They'd largely lost touch, more through Jay's fault than his. He'd visited when he'd been in medic school, but it was ages ago. The last he'd heard, Sutter saw the VA for a back injury he'd gotten overseas, and he'd married another soldier he met in the National Guard. They had a daughter together.

Jay stared at the words, paralyzed. It worsened when he saw the time stamp on the text. Sutter had sent it an hour ago.

Jay leapt out of bed, sending the lamp on the bedside table crashing to the floor.

"No, no, no."

With a shaking hand, he hit the Call Contact button above Sutter's name. He paced back and forth as it rang, his dinner from the night before threatening to come up. His old teammate, his roommate and friend of ten years, was already dead. He'd killed himself an hour before after reaching out for help, wondering why nobody had answered his cries.

Tears formed in Jay's eyes as the call switched to voicemail. "No, no, no!"

He punched the wall. *Boom!* A fist-sized hole appeared, and he pulled his hand out of it covered in dust, blood, and drywall. He hit redial.

While it rang, his thoughts gathered into a shivering mess. Sutter's address might be on Facebook. The VA would have his number. The local police would know how to get ahold of him. It might not be too late yet, as long as Jay could find a way to get someone there.

He logged into his laptop as the call went to voicemail once more. He typed frantically and brought up Sutter's Facebook page. He'd moved to Arizona at some point. Despite the adrenaline surging through Jay's veins as he hit redial once more, he felt a pang of regret at seeing that. If anyone had said eight years earlier he'd lose contact with any of the old mortars, he would have told them they were out of their damn mind. They were family—more than family. His own parents meant less to him than the people he'd bled with.

And now Sutter was dead, lying in a pool of his own blood, his wrists slashed. He was leaning against wood paneled walls in a small apartment, a hole in his head where the bullet had ripped through. He was—

No. Jay shut the thoughts down. They weren't going to help.

289

Pine Top, Arizona. When the call didn't go through again, he called the police there.

"Pine Top police, is this an emergency?"

"Yes. I have a friend I think is…" He swallowed past the lump in his throat. "I think he's about to commit suicide. I'm in Tennessee. I can't get to him."

"Okay, sir. What is his name and address? Is he on the other line?"

Jay couldn't stop pacing. "No. I can't reach him, but he sent me a message. I don't have his address, but his name is Jerimiah Sutter."

"One moment, sir." Fingers flew across keys with a *tick, tick, tick.* "I don't have an address for him. Can you get ahold of him to get it?"

Jay's mind raced. "No. But I might be able to through the VA. Let me call them, and I'll call you right back." He jammed the wrong button twice before ending the call, his fingers shaking. He looked up the number and dialed them. Their endless automated line wasn't cutting it today, and he jammed the zero on his dial pad until a human voice answered.

"Phoenix VA, how can I direct your call?"

In one long breath, he explained what was happening.

"I…um. Hold, please." The woman sounded unconcerned. Not scared for a veteran, just worried she might be on the wrong end of a scandal in a day's time.

A man picked up next. "Phoenix VA, security."

Again, he explained it to the man, doing his best to hold back tears.

"Slow down, sir. Is your friend already dead?"

The words sent his blood pressure through the roof. "I don't know! Are you fucking listening?"

"I'm trying to, sir." The man didn't sound the least bit put off by the tone. Unlike the woman before, he actually seemed worried. "Do you have any way of getting in contact with him? I'm a veteran too. I get it."

Jay's anger ebbed a little, but not much. "No. I just need his address. I can get the police to his house if I can have that."

The man blew into the phone as if every second wasn't precious. "I can't give you that, sir. It's a matter of patient confidentiality."

Jay's mouth hung open as he stopped dead in the hall. "Are you fucking kidding me?"

"No, sir. We can't give out patient information."

Any semblance of control vanished. "You listen to me. If my friend dies because you were too scared to do anything, I swear to god I'll find you and kill you, do you understand me? I'll cut your fucking head off."

Silence greeted him. Jay stared at the wall ahead, but he was somewhere far away. A place where one of his closest friends bled to death with a knife in his hands.

When the man spoke again, his voice remained calm. "Okay. I'll call the police and give them his address."

Jay closed his eyes. "Thank you. Please, hurry." He poked the disconnect button and redialed the police. Before the woman could even speak, he told her what had just happened. "Whoever goes, please, make sure he knows Sutter isn't dangerous. He's just scared and alone. I don't know what's happening, but please, send another veteran if you have one."

"I will, sir."

"And call me as soon as you hear anything, please."

He hung up, but as soon as he did, the phone rang. It was Sutter. "Jerry?"

Silence for a moment, and then he heard Sutter's voice for the first time in six years. "Hey, man."

Relief flooded him in a way he'd never known. Even when they'd survived Najaf without losing a single man, it had never felt so complete. "Jesus Christ."

"Sorry." He said it as if he deserved punishment for something.

Jay's heart broke. "No. No, don't be. Shit, man. What's going on?"

Sutter cleared his throat and sniffled. "I'm all fucked up. The VA had me on pain pills for my back, and then they took me off when they found out I was smoking pot. I went from three Percocet a day to nothing in the time it took the cop waiting outside the office to walk me to my car."

Jay balled his hand into a fist, but he tried not to let it show in his voice. "I'm sorry, man. But you don't need to be. I'm glad you called me before…before anything happened." It was just like the VA to fuck something up like that. Put someone on something they couldn't live without and then pull the rug out from under them.

"I'm watching my little girl. I couldn't do anything with her in the house. I just… I don't know what to do. I've been using all our money to get pain pills. Mary doesn't know. I haven't worked in years. I-I'm a fucking mess. I'm a loser."

The way his words mirrored Jay's feelings was more than he could handle. To hear even him speak like this was horrifying. This man had asked after every mission if Jay had killed anyone. The guy who had proudly spoken of standing up after Nine-Eleven to fight for the country that had done so much for him.

"You aren't a loser. No fucking way. Losers don't do the things you've done, and losers don't ask for help. They certainly don't put their kids first when they're at rock bottom. You're a winner, J. J."

Sutter laughed, and for a moment, he sounded like his old self. J. J. had been a running joke in the squad. One night while drunk, Sutter had told them how his mother had named him Jerimiah Joseph Sutter just so she had a son she could call J. J. The only problem was she forgot to call him that until he was a teenager.

Sutter sighed into the receiver, and the new, broken man returned. "I didn't know who else to turn to man. I couldn't go to Mary. The VA can't help. I'm just… I'm fucked, man."

Jay couldn't recall ever hearing a man so tired. "You aren't fucked. And if you need me, I'll be there in the time it takes to put pants on and get on a plane. You aren't alone, brother. I'm here. We're all here."

"Yeah?"

"I've been on disability for almost a year now. I've hit some deep lows myself. You aren't alone. I swear it. We're going to get through this."

The scrape of a chair pulled across hardwood screeched through the line, and for one insane moment, he thought Sutter was about to hang himself. Instead of the sound of a rope snapping, there was only a man sitting. "I find that hard to imagine. Your ass was always too optimistic for that shit."

Had he been? He'd been the youngest in a group of twenty-something killers, most of whom would go out drinking on the weekends when Benjamin hadn't been old enough. Donner had once remarked that Jay wasn't scared of anything. Not of getting smoked, not of getting hurt—nothing.

But that was a long time ago. "I swear it. I can't imagine what you're going through with the withdrawals, but you can bet your ass I'm going to help you through it." He recalled the police on the way to Jerry's house. "Speaking of which, I, uh… I got your address from the VA. I had them give it to the police so they can come and check on you. Don't freak out if they show up, okay?"

"You did?" Sutter didn't sound angry or hurt.

"Yeah, but they're just coming to check on you."

"It's fine. It's just me and Piper. Mary is at work. She… Man, she's going to find out I spent all our money. She's going to think I'm a fucking loser."

Jay stepped out onto the porch and lit a cigarette. The cold bit deep without a shirt on, and people driving by the apartment gave him odd looks, but if he didn't do something to calm his nerves, he was going to explode. One wrong step might put his buddy over the edge.

"She married you, dude. She doesn't think you're a loser. Whatever you're going through, she can help you." He thought about it as he stared into the copse of trees across the street. "Unless she's an asshole."

Jerry laughed once more. "She isn't. I know you're right. It's just. It's hard to feel like anything matters anymore, you know?"

He did know. He knew exactly. "Sometimes victory means just getting through the day, man."

A knock sounded at the door on Sutter's end.

"I think that's the cops. I'll call you right back."

"You call me back, man. I mean it. Call me and we'll make a game plan. We'll figure this out together, brother. Just like the old days. I love you."

"I love you, too."

It took longer for the police to leave than Jay expected. They refused to go until Sutter called his wife, but true to her word, the dispatcher had sent a veteran. The man had apparently been through something similar, and he made sure to take good care of Jerry.

Mary had been mad, of course, but she'd been much more understanding than Jerry had assumed. That was the rub of fatalistic thinking—the assumption that the worst thing that could happen would always happen. Every good thing you touched, every good thought you had, was nothing more than a bump on the road to more misery.

Maybe it was true. Jay still lived too deep in his own bad times to figure it out. But having something outside himself to focus on helped, and being there for his friend made a world of difference. He'd almost tried to articulate how much it meant that Jerry had come to him in his darkest hour as they spoke often over the next few weeks, but he couldn't. Words wouldn't be able to sum up his feelings, and it wasn't about him anyway.

Within a week, Jerry went to rehab through the VA. Jay gave them what money he could spare under the condition they never think of it again.

"You're family, and it's only money. I'd give you all of mine and steal as much from the next guy to make sure you're good, brother."

The VA made promises up and down that they'd fix it so that didn't need to happen, but they never apologized for their fuckup.

Why would they? In the modern era, nothing was ever anybody's fault. They could pass the buck forever. Responsibility was a word often used and seldom embraced.

When it was all over, he could breathe again. Jerry was on the path to recovery, and Jay was back in contact with a man he should never have lost it with in the first place. The crisis had been averted.

Jay had never approached his friends for help. He'd never bothered to cry out like Jerry had, content instead to let his people suffer as much as he was if he went through with it. The weakness of character in that was uncomfortable to look at.

It had been a rough month, and he'd helped his friend through the worst of it. If nothing else, he could hold onto that.

Chapter 27

The shit-water rose to their knees. The sewer system in the neighborhood had long since been destroyed, and the whole place reeked of human waste and garbage. Benjamin had seen nothing worse in his year at war.

Even before the war, no way had this ever been a paradise. Oppressive summer heat notwithstanding, the buildings were one step away from rubble. Cracked stone, paint long gone, roofs bare to the supports in some places. Piles of refuse as tall as the children playing in it stood wherever there wasn't shit water. Bullet holes covered everything. The blows of unseen explosions pocked stone fences.

But today, none of that occupied his attention.

"Blue six, black six. Go ahead and back up now. I think you guys have it this time. Over."

Benjamin watched through the periscope as the third platoon Stryker inched forward, pulling the remains of the Humvee out of the sewage-filled crater.

It had started as a joint operation with another company from another unit, a knock-and-clear of a neighborhood known to harbor terrorists. Hundreds of pounds of explosives had detonated under the street. At the best of times, this area was incredibly dangerous, and as cold as it was in light of the tragedy only an hour old, he thanked his lucky stars they were in Strykers.

The Humvee never stood a chance, and everyone inside had died instantly. Rosenberg peeled the driver off the steering wheel. Another member of the crew had smashed halfway through the bulletproof windshield. Everything inside was a tangle of blood and body parts.

The other unit had left shortly after. Military policy dictated you not recover your own people if you could help it. War scarred men enough without staring at the mangled remains of what had once been your best friends. Even though he hadn't known those people, it has hard for him to look. They were distant relations, in a way. All members of the same team.

All of them gone before their time.

The recovery took most of the day because of the sewers and narrow streets. Benjamin stared as Rosenberg and the others picked up chunks of young soldiers and placed them in body bags. He finally looked away as his friends tried to untangle the corpse of a teenager from the gunner's turret of the Humvee.

Iraqis watched from rooms above or from shaded alcoves on the street, and Benjamin tried his best to hate them. How many had known the bomb was there or had taken part in placing it? How many of them would go to sleep tonight saying prayers of thanks to Allah for another group of infidels killed?

Despite his best efforts to work up some emotion other than resignation, he was unable. All he wanted was to go home. To see himself and all the soldiers in that Humvee back on American soil and away from this shit-covered street in this shit-covered country.

By the time they returned to BIAP, night had fallen. There would be no missions the rest of the evening, a small miracle. Or at least no missions were planned. Terrible things had a way of cropping up around here, and now that they were getting into the meat of the extension, it only grew worse.

He trudged to the shower trailers to wash another day of filth and sweat off. The PT shirt he wore to the trailer had faded from too much time in the sun. All of his uniforms showed signs of wear. All of his friends too.

When he walked inside, Sergeant George Smith was there. Smith had been with Kennedy, Jared, and Benjamin in basic training, but he hadn't been a new solider. He'd been a POG of some kind or another before deciding he wanted to be a real soldier. Benjamin cringed at how little respect the guy had gotten despite already being a veteran. New soldiers had no concept of what that meant.

He'd ended up in Alpha Company of the Patriot Battalion, only a few buildings up from Benjamin and Jared. Still, they seldom saw one another.

"Hey, Sergeant Smith. How's it going, man?"

Smith stepped out of the shower. The answer was as clear as the bags under his eyes and the worry lines that hadn't been there when they'd met three years before.

"It's going."

Benjamin shed his clothing and turned the shower on. The trailers might have been nice at one point, but they'd seen years of war and thousands of feet. The worn tile floor sagged beneath them. Benjamin found it hard to imagine being at war as long as they had without a shower, though this was one of the first wars where such a luxury was a given.

Smith shaved in the mirror while Benjamin waited for the water to warm.

"You know, I saw Kennedy a little while back." He couldn't remember if it had been weeks or months. Everything blended into a steaming bowl of shit soup here. "Whenever we were down at Kalsu. Guess that was a while ago now."

Smith's hand holding the razor froze. He stared at Benjamin in the mirror as if waiting to hear a punch line.

"He was an E5, though I guess that's not the least bit surprising. Guy is going to be sergeant major of the army one day."

Smith faced him directly while Benjamin grabbed his shower kit and stepped under the water.

"You didn't hear?"

Benjamin turned to him. "No. What?"

"Kennedy died."

The kit fell from his hand and thumped onto the floor. He hadn't meant to be so dramatic; it was an involuntary reaction. It took a second for him to recognize the joke, and Benjamin smiled. "Dude, I just saw him. That's fucked up."

But Smith wasn't smiling, and he didn't respond.

Benjamin, standing half in the shower and staring at his friend of three years, grinned regardless. He'd just seen Kennedy. It had only been a few…

The expression on Smith's face broke through. Not anger. Not pity. Not looking down on Benjamin for needing a moment to see the horrible truth. Just commiseration.

A sharp pain tore through Benjamin's heart, a ripping he had never felt before. Every infantry instinct in his body fought it, but his eyes began to water. "What?"

Smith nodded before returning his gaze to the mirror. "He was killed by an IED in April."

He didn't follow it up with, "I'm sorry," or any of the other trite nonsense people said after shattering your world. He didn't need to.

They dealt death on a daily basis. It was a visitor as often seen in the infantry world as rain and misery. Sorry was a given when two men from that environment lost a third.

"Are you sure?" He blinked back the tears in his eyes. It stung to do so. Without getting in, he shut the water off and put his shorts back on.

"Yeah, man. I'm sure."

Benjamin quickly put his shirt on and picked up his bag. He mumbled, "Thanks," before charging out the door.

Smith was smart, and a good NCO, but even he could make mistakes. Benjamin's heart lightened as he thought about it, dashing back toward his CHU to grab his rifle and shoes. It was an error, nothing more. Smith had heard about a guy named Kennedy getting killed and confused the two. Benjamin had just seen the man. No way he'd died.

Kennedy's words drifted to Benjamin across time, back from the days when he was running a platoon and keeping a stupid teenager in the army who'd done nothing but give him shit. "You can't sweat the things you can't change, man. You just have to accept them and move on."

His stomach churned, and his lunch threatened to come up. He hated that weakness. Kennedy had taught him better than that. He *was* better than that. Nobody in this company was weak, and he wouldn't be the first.

"I need to go to the MWR," he said to Sutter as he entered the CHU. "I need to check on something. Right now."

Sutter had one earbud in as he stared at his laptop. He pulled it out with an exasperated sigh and looked up. "What, bitch? We just got back from—" He appeared to notice Benjamin for the first time.

Benjamin didn't know what expression he wore to make Sutter stare at him like that. "Please, man. It won't take long."

Sutter sighed again before rolling off his bed. "This shit better not take all night, dude."

Benjamin grabbed his gear and a cigarette before he stepped outside to wait for Sutter. He lit the smoke and paced back and forth as he puffed away at it. A moment later, Sutter joined him, and they trekked to the MWR. It wasn't far, but he had to stop himself from running. Only a fool ran for something so mundane. A mistake. An accident. In twenty minutes, he'd be tracking down Smith to correct him.

"Did something happen?" Sutter asked.

"No. I just need to check something." And that was the truth.

"Did Biffany break up with you? Do you need to put some poetry on Myspace?"

Benjamin ignored him. They reached the MWR, but there was a line for the computers. They sat next to one another without saying a word. All around them, soldiers tried to forget the war for a little while. AFN played on a nearby TV. The computers were awash in people laughing while watching videos or looking at pictures of their friends and family.

That certainty he'd felt on the way over drained as he sat there. He'd thought it was a joke when Henkes and Griff died too. He'd been wrong every time. Too stupid to see the writing on the wall. Gallows humor hit big among soldiers, but this wasn't a laughing matter. And Smith…that emptiness on his face.

Someone called Benjamin's number, and he strode to the computer. He pulled up Google and typed in Kennedy's name, but his finger hovered over the mouse, hesitant to press search. This wasn't going to give him any closure if it was true, and if Smith was wrong, he could find out in a few months when they got home. No good would come of this.

But he had to know. He clicked the button.

The first result popped up immediately. An obituary from Kennedy's hometown in Maryland.

Benjamin stared at the page as the bottom fell out of the world. People still laughed and talked. An AFN commercial implored soldiers to be careful about shaking their babies. Any other time it would have gotten a smile out of Benjamin, but there were no smiles now. For Kennedy, there would never be smiles again.

He stood up, nearly knocking over his chair. He'd seen death before. In this place, they contended with it daily. He'd lost friends and teammates. But this?

He walked back to Sutter. No tears threatened this time. The enormity of the news wouldn't allow it.

"We can go."

Sutter's earlier annoyance vanished. He stared at Benjamin with genuine concern. "Jesus, dude. You look like your dog died. What happened?"

Benjamin turned to walk out the door with Sutter in tow. "A friend of mine was killed."

The sun painted the desert in deep reds as it set. Blood across the canvas of the world. Soldiers went about their day, some on their way to the nearby chow hall, others to the PX. Benjamin walked back toward the CHU and gazed at the sparsely clouded sky. He scanned the horizon, staring at the space where the nearby road curved behind T-barriers and vanished. He turned his sight anywhere but inward. He couldn't. If he did now, he wouldn't be able to go on the next mission. Always one ride away from the bomb that would take your life on some dusty road for some mission forgotten within a week.

"Are you going to be all right, man?" Sutter asked.

No. Nothing would be all right. Kennedy's family was probably still visiting their son's grave every day, cursing themselves for instilling in him that driving need to take up arms for his country. Or maybe they had told him all along this was how it would end, and he went anyway. Kennedy was that kind of guy. He *had been* that kind of guy. Now he'd died, and the world was less for his passing.

"I'm fine."

Chapter 28

Jared's visit had been planned for months. After they'd all gotten out at the same time, Jared had stayed near Ft. Lewis in Washington instead of going back to Mississippi. The last time Jay had seen Jared was the day Tiffany had broken up with him. She'd flown out to Washington to visit for a whirlwind weekend of sex, drinks, and good times. They'd seen every sight, though the sight they saw more than any other was each other. Those first few months back felt like a dream now.

It would be impossible to explain to anyone who hadn't experienced it what it had been like, to go from being mortared on BIAP one day to shopping for pants in Tacoma the next. Flirting with the girl at the cell phone shop. Eating barbeque until his stomach hurt and laughing about it in a drunken stupor. That night, he and Howard had drunk-dialed everyone on their chain of command list, laughing at the groggy sergeant majors and colonels as they answered their phones.

What could better encapsulate the shell America lived in? What better way to show the citizens what it meant to be on the outside? And yes, it was bizarre and wonderful. To go from being a warrior to not. To be a hero among mortals. But hero faded, and wonderful did too. Soon only the strangeness remained, and then only the anger. The disappointment. The nagging feeling it was all for nothing.

But those nagging feelings hadn't been there on those nights with Tiffany in Seattle. Some people had only a bottle of liquor and a bunk in a rotten, old building. But not him. He might have the same bed he'd had before deployment, and he shared that liquor, but he had someone waiting for him at the airport the following weekend.

But as that weekend had come to a close, she dropped the bombshell. She couldn't be with him anymore. They'd drifted apart. She needed space.

He'd thought about leaving her at the hotel instead of driving her to the airport, but that was petty, and he was on cloud nine anyway. He didn't need her. He'd survived fifteen months in hell, and the world belonged to him now.

Jared, Allen, and Jay had all left the company for the last time as a group a few days later. Laughing as they went their separate ways. Allen and Jay had known they would see one another again in a few weeks, but neither knew when they'd hear from Jared again.

And now they did. To think his friend would visit him again tomorrow made him both giddy and nervous. Nine years ago, he wouldn't have been, but things had changed. He wasn't the carefree twenty-one-year-old he had been. The weight of the things they'd done together dragged him down in a way he couldn't have imagined when he stepped off that plane in Washington.

When he was eighteen and thought he was saving the world, it didn't bother him. But people got older, and if they were lucky, wiser. With wisdom came doubts. Had they the right play? Did they make it all worse? Was all the blood that came after on their hands?

He talked to Howard about it online that night.

"It'll be fun, dude," Howard said once Jay had admitted he was nervous. "It's fucking Jared. What, do you think he's going to beat you up?"

Jay laughed as he sat on the couch in the apartment. Jared was one of the nicest guys he'd ever known. "No. I don't know, man. I just get nervous about stupid shit. Why don't you come up too?"

Howard only lived a few hours away in Alabama. "I dunno. I have a lot going on this weekend. I don't think I'll be able to just drop everything and stop by."

"Well, man, if you can, just let me know. I'm here."

"I will. What's your address?"

Jay gave it to him. They spoke a little longer before Jay made excuses and left the chat. He didn't have the energy to speak to others for long periods. He was getting better at learning the limitations of his illness through the VA, but it was still hard to accept something was wrong when he couldn't see it.

Still, as he closed his eyes to sleep that night, excitement more than fear kept him awake. It might mean things were getting better, or it might not, but he smiled about it all the same.

The next day, he dressed to meet Allen and pick up Jared when there was a knock on the door. He spit the mouth full of toothpaste out and threw a shirt on as he hustled to answer it.

He didn't know what he expected, but it certainly wasn't Howard.

He stared at Jay with the dead-serious expression that meant he was about to tell a joke or throw a punch. "This is what you do with your life? You look like you just woke up."

Jay shook his head as Howard's smile burst forth, and he embraced his friend. "What the hell are you doing here? I thought you couldn't make it."

"I lied." Howard stepped inside. "Did you really think I was going to miss a chance to see all three of you guys?"

Jay's mind wasn't quite catching up to his friend, a decade absent, being in his living room. He stared stupidly before excusing himself to finish getting ready, shouting over his shoulder, "Give me a minute and we can run by Allen's house. It'll be one hell of a surprise."

They both laughed, but Jay needed a moment alone to process it. Twenty-one-year-old him had loved surprises like this. Now, he stared into his own eyes in the bathroom mirror and reminded himself he just needed to breathe. One in, one out. Keep on like that until things calm down a bit. Don't be afraid to press the pause button on a situation, and learn to be okay with not being okay. All the lessons his doctors at the VA had taught him came to mind, and slowly his calm returned.

A few minutes later, he stepped back into the living room, and his smile returned. "You ready, buddy?"

As simple as that, things fell back into that old pattern. Laughing, joking. Catching up on times now long past but still fresh to a friend not seen in ages. They drove to Allen's. He wasn't home, so they waited on his front porch, smoking and bullshitting.

"You look exactly the same," Jay said.

Howard shrugged. "I don't feel the same. The last few years have been long."

"I hear that noise." Jay flicked his butt into the grass and watched the smoke curl up.

"I got married, and then she fucked my neighbor."

Jay raised an eyebrow. "Really? That's some *Maury* shit."

Howard smiled and shook his head. "Yeah, you aren't fucking kidding. She didn't even get anything out of it, either. The least that asshole could have done was get her pregnant so we could get some child support. I had to dump her after that."

Jay slapped Howard on the shoulder as Allen pulled into the driveway. When he caught sight of Howard as he climbed out of his car, he clapped. "It's just like you to show up out of the blue."

Howard did a small bow. "What can I say, man? You have to keep it interesting."

Allen hugged him as he stepped onto the porch. "Couldn't resist the chance to get the band back together, huh?"

"We're missing about a hundred and twenty five guys," Jay said.

Allen laughed as he let them inside. They drank and talked, ordered pizza, and reminisced. That night, they drove to the airport to pick up Jared. They leaned against the railing outside the single security line, waving when he exited a gate at a distance. He hadn't aged a day either.

Jared's reaction was much the same as Allen's upon seeing Howard.

"Oh, what the hell! Who let this guy out of his kennel!"

The anxiety vanished as they stood around talking to one another at the baggage claim, and Jay felt ridiculous for ever having experienced it in the first place. Nothing felt more like home than a friend, and no friend was as close as those you've been to war with.

"So, you became a cop?" Howard asked Jared on the drive back. "I could see that."

Jared glued his gaze to the window and took in the sights as the car sped along the highway in the waning light. "It felt like the right choice. It was an easy transition from the military, and I still get to help people."

His country accent had vanished in the years since Jay had seen him last. "You're not out there wrecking house?"

Jared laughed. "Nah. I play basketball with the kids on the poorer side of town every week."

That was just like him.

They pulled into Allen's house, and the drinking commenced in earnest. By the time they were two dozen beers deep, it had all fallen back into a comfortable rhythm. The old jokes and stories came back to life. The time Howard had taken a picture next to a blood stain on a wall, pretending to be dead and sending it back to his family as a joke. The time Benjamin and Donner had set off the fire alarm in the barracks during a fight with a can of foot powder. Stories of their friends and the places they were now. Allen and Jay had stopped talking about those things years ago. They'd been friends for over a decade, and it had become old hat to them. Just another set of memories in a long line of them.

It was good to revisit the past, but as it always did, the stories took a turn. Jay spoke of his disability, not making eye contact as he did. Howard talked about divorce, and the failing coal industry. They all shared stories of the dissatisfaction with the world, the lingering feeling inside that all wasn't as it was supposed to be. They were deep in their cups by the time Bales came up.

"They should have put a bullet in him," Jay said. "After what he did? I don't know how he could stand to look at himself in a mirror. It would be a mercy."

Howard shook his head. His jaw set. He looked every bit the killer he had been a decade before. "How would you feel if you got thrown in jail and your family turned their back on you like that?"

Jay shrugged. "I wouldn't be in that situation to begin with. We weren't murderers. That wasn't what we did."

Howard glanced away, taking another drink from his beer and locking his eyes on the wall.

"I don't know," Allen said. "I can see both sides. But ever since I heard about it, I can't help but imagine how I would feel if they were my kids."

He'd stolen the words right out of Jay's mouth.

Jared sat on the floor with his legs crossed. He'd been quiet most of the evening, but he finally spoke up. "There's no going back from something like that."

Jay got it. He wasn't so far removed from war that he couldn't understand seeing everyone outside of the uniform as an enemy. Civilians. Politicians. Iraqis. Afghanis. All were just one bad day away from being on the wrong side of the barrel. But that didn't excuse it. Nothing did.

"We all have our lows, but fuck him. He can rot in prison for what he did. He isn't that different from the asshole that killed the five guys in Chattanooga earlier this summer. The only difference is that guy wasn't on the team we were rooting for."

If he hadn't been drinking, he wouldn't have spoken about it at all. He instantly regretted having said anything. It was better to keep his mouth shut. Why ruin a reunion like this with painful memories?

Howard shook his head, and his expression softened. "But he's family. It don't matter what he did. He doesn't stop being family when he fucks up."

That got through to Jay, and he stood staring into his drink for a long moment. Family. No matter what. No matter how far they went, what they did. No matter if they were all together, or if he was the last one left alive, just waiting to pass on and join them in whatever came after.

For the first time, some small measure of pity stirred when he thought of Bales. Not as strong as the pity for his victims, but pity nonetheless.

"I've been to some low places, too." Jared stared at the carpet as he spoke. "I thought about leaving my wife and kids and just driving away a few times. You know, just get the hell out of Dodge and start over somewhere."

A quiet settled over the room, the calm that followed things between friends and needed no words. Nothing would fix it. Not Jared and not Jay.

When Howard spoke again, he did so without shame. "I tried to kill myself last year."

All eyes shifted to him.

"I was…I was just done. I got in my truck and rode over to a bridge that I was going to just drive off of. I was a little drunk, a little down, and I just didn't see the damn point anymore. But right as I got on the bridge, my timing belt broke." He laughed. "I had just changed the son of a bitch the week before. But my truck stalled right as I started speeding up. When I had a minute there on the bridge to think about it, I decided not to do it. If that doesn't make you believe in God, I don't know what will."

To hear Howard's low place so closely mirrored his own was disturbing. On most days, he was just a little bummed out, a little anxious. On others, it felt as if there were no coming back. As if all the good things he'd ever done meant nothing, and all the bad was being counted and stacked. As if the world was waiting for the opportunity to level its evidence against him.

But it was all about dissatisfaction with the world, even if misdirected. The good and the bad. Once a man had seen the darkest parts of humanity, if it didn't infect him, he felt a calling to make it better. And if that calling were to fail him, would he end up on the streets? Drugs? So full of hatred and self-loathing at his failures he couldn't stand it? No. They never stopped being warriors, even if their deeds had sometimes been misguided, even if they had long since taken off the uniform.

But it was harder to live for something than to die for it. That had been why he didn't think he would come home when they were

317

in Iraq. Why there were days he just wanted to take the easy way out.

"I'm sorry, man," Jay said. "And I'm glad your goofy ass is still with us."

"Yeah, me too." Howard took another sip of his beer. "But don't feel bad about taking money if you need it, man. I know you, and I know you do. That's what the money is there for. There's nothing to be ashamed about. And you never know. There might be someone out there watching you who isn't as strong. If they see you getting bent out of shape about it, how do you think that would make them feel?"

Jay knew Allen had hit low points too, but he kept silent. For some people, the best way to deal with it was quiet.

That was enough somber talk for one evening. Jay polished off his beer in one swig. "I say we stow this shit and head down to Waffle House to grab some food we'll regret in the morning. Who's down?"

They all made their less-than-sober way to the car, laughing and joking just like the old days. Thoughts of suicide, murderous teammates, war, and patriotism were gone for a while, drowned out in smiles and warm, far too infrequent moments.

Jay drove home that night and fell into the kind of terrible sleep reserved for those who drank too much and cared too little. When he woke the next morning, he called to find Howard had gone back to

Alabama. Jared and Allen would be seeing the tourist traps that day, but Jay had other plans. He excused himself and opened up his laptop.

He wanted to say something about everything that had come to pass. There were other books on the subject of war, PTSD, and a soldier's place in the world, but that didn't mean the last word had been spoken. Hearing Howard's story the night before, talking about Bales... Their story, and the story of their generation of soldiers, needed to be told from the ground level. Not to try to justify the war, not to show what it had been like on the battlefield, but so that future generations would be able to see who they were. Maybe nobody would read it, and that was fine too. He was doing it as much for himself as posterity.

His first thought as he opened a Word document and began typing notes was, "What would Kennedy think?" He'd been dead so long now nothing but bones and cloth remained in his casket. Jay had been nothing more than a scared child, but Kennedy had been a man. He'd known what he was getting into. What it meant to serve. So he'd signed up, he'd gone to war, and he'd been blown to shit in some rusty Humvee on some nowhere street. Another victim of the meat machine. Another faceless pawn turned corpse to be ground up and spit out for the corporate agenda in a war of lies.

They'd been called heroes when they killed. When people had stopped saying it, it faded, and they went back to being mere men once more. Maybe both were true. Maybe neither. Maybe nobody cared because nobody should. Maybe the very idea of a hero is the stuff of fairy tales, not humanity. Not reality. There were no heroes, just men and woman willing to sacrifice. That was as close as the real world ever came.

Before he'd left for Iraq, he'd thought sacrifice meant going without comfort. Being away from everything you loved. Not anymore. Now he knew better. Personal sacrifice meant cutting off a piece of yourself, you and all your brothers, and leaving it on some ancient road in some backwards city on the other side of the planet. Sacrifice meant losing the ability to unsee how pedestrian the cost of a life could be. Being unable to pull the wool back over your eyes. Talk might be cheap, but it wasn't half as cheap as blood.

But that didn't have to be the end of their story. Not him. Not Henkes or Griffin. Not Kennedy. It was amazing how quickly the world forgot tragedy. Three thousand dead in New York. A million natives in the Middle East. Ten thousand Americans at war. Five in Chattanooga. The next day it was all past tense. Life went on even for those closely affected, and within a few years, it was barely remembered numbers and voices.

That wasn't how he was going to live. He'd remember them. All of them. He'd make sure they lived forever.

Chapter 29

It wasn't often that he spoke of missions in his letters to Tiffany, but tonight it felt important. A pall hung over the day. Something slithered in the air—a malice or discomfort that spoke of bad things to come. And so he wrote about it to her, telling her he loved her and that everything would be okay no matter what happened to him.

It was depressing fare, but he finished it and sent it off before the mission that night.

Nothing about it was out of the ordinary, just a patrol in the bad section of a town that didn't have a good one. They'd been doing the same thing for…thirteen months? Fourteen? If he felt like ruining his day, he could count it.

They sat around the Strykers waiting for SP that night. Simmons and Masters argued back and forth. Sutter smoked on the edge of the ramp while Olson told him stories of the things his wife and child were doing back home. Garcia, Miller, and Brenner stood in a circle nearby, discussing the mission. Benjamin looked up at the streetlights around their parking lot, the dread in his gut growing.

Nelson, who'd been leaning against the Stryker nearby, kicked him. "What's your problem?"

Benjamin tore his gaze from the light and shrugged at Nelson. "Just got a bad feeling about tonight."

Miller had banned Benjamin from using the term "bad feeling" after he'd abused it in Mosul. Nelson kicked him again. "Well, fuckin' don't."

They loaded up when the time came and drove off BIAP, weaving through barriers and past checkpoint after checkpoint. Men and women in uniform waved them through while people on fifty-cals eyed them warily. In a moment, they were on route Irish, making their way toward their patrol route.

Duarte and Olson picked at each other on the headset as Brenner directed him. At night, the black and white DVE the drivers used didn't do as much as it could have, and so a TC had to give constant directions. About thirty minutes out of the gate, they turned down the road they were looking for.

Benjamin's stomach dropped as he stared at it.

"Go forward. Pick up some speed," Brenner said.

"Will both of you jackasses shut up?" Nelson said to Duarte and Olson in the back.

Benjamin hit the accelerator. The Iraqi police station on their right stayed lit up like a Christmas tree all night, but everything else was black. Not a single living soul wandered the streets. Everyone had to be in before sundown or risk their lives.

That unease in his gut grew into a physical sensation, something reaching up from under his seat and pulling him down. Something appeared on the DVE just in front of the vehicle. Real enough to leave a little white box on the screen, but not close enough to make out any detail. A piece of rock, maybe. It didn't look like something off the road but part of it.

He adjusted at the last second, but it was too late.

Boom!

What he was seeing made no sense. All the lights in the Stryker had gone out. Smoke. His left earbud remained in, blasting rock music, but he didn't need the distraction right now. He tore it out.

Someone yelled in the back, but he couldn't tell where that was. Every warning alarm he possessed screamed at him. People were hurt. His insides scrunched up as his adrenaline sang. His head ached.

He touched his face. Oil covered his hands but no blood.

"Benjamin! Benjamin!"

Someone had been shouting his name, but he hadn't noticed past the ringing in his ear. The constant roar of the engine and hum of the electronics vanished, replaced only with yells and ringing silence.

"Benjamin!"

It hit him. An IED in the road had blown them up. His eyes widened and his mouth dropped open. The headache raged as he grabbed the flashlight off his belt and directed it into the hellhole behind him.

A giant hole gaped straight through the bottom of the truck right behind his seat and just in front of Brenner a few feet back on the other side. Black smoke and dust filled the air, making it difficult to see, but that was clear as day. If it had been even a few inches in either direction…

"Benjamin! Answer me! I can see your light. Do you need help?"

It was Brenner. Behind him, he could vaguely hear the others.

"I'm okay!" Everything hurt, his head more than anything. Had it struck something? He couldn't recall. "Are you guys all alive back there?"

"Yes! Get out! The rest of the platoon is pulling security!"

He climbed out of his driver's hatch and nearly tumbled off the Stryker getting down. Howard and Jared pulled security at the front of the vehicle as he fell more than stepped off the slat armor.

"Jesus fucking Christ." Benjamin untangled himself from his weapon sling. The whole world shook, or maybe he did.

Howard grabbed him by the shoulder. "Go see Pac."

Of course they'd want him to see the medic after something like that. He looked up at his Stryker as he walked to the tanker vehicle where Pac waited. Big, silent, and inert. The thing that had protected him his entire deployment lay as dead as the ground under it.

That thought shook him more than it should have as he climbed into the other Stryker.

Pac felt Benjamin's neck and checked his head for any signs of injury, but Benjamin wasn't watching at him. He watched something far away that even he couldn't pinpoint.

The radio buzzed with people speaking back and forth, and it took a moment for Benjamin to catch up. "Green 6, Red 6. We're leaving now. Over."

"What's red platoon doing here?" Benjamin asked as Pac waved a light in front of his eyes. It left trails in his vision where it passed, and he knew enough about medicine to understand that wasn't a good sign.

"Nelson was hurt."

The word rocked him back in his seat. "What! Where is he?"

Pac grabbed his shoulder. If he meant it as a comforting gesture, he failed. "He's on his way to the hospital. He's okay. It just knocked him out."

The military talked all the time about the dangers of repeated exposure to explosions and head injuries, and Nelson had more than his share of both. Benjamin stared past Pac at the panels behind him.

It had finally happened, just like he knew it would. They'd been hit. By some miracle, nobody had died. "Is anyone else hurt?"

Pac shook his head. "The lieutenant got some shrapnel, but he's fine. It looks like you took a hit to the head too." Pac stuck him in the back of the tanker vehicle and told him to stay there while the rest of the platoon hitched up the broken Stryker to recover it. He felt like a child. His friends were hurt, and he was put away.

Not long after, they began driving back toward BIAP.

Garcia ducked out of his hatch and pulled back his CVC. "You okay?"

The motion of the vehicle spun his head. "I think I'm going to be sick, Sergeant."

"Okay. Don't do it in here."

He switched places with Ski in the back hatch and puked off the rear of the vehicle as they sped through the night. It tasted of stomach acid, and it dripped down the armor as it oozed out of him.

He didn't feel better, and when he was done, he stared into the night. There were no lights, and without his NODs on, the world became a starry void. He gazed at those million pins of light stuck through the black cloth of the universe. It was so easy to forget they were the same ones you saw back at home. That somewhere in the world, someone looked up at the same stars, thinking thoughts as different from his as summer from winter.

He sat back down in the vehicle. Garcia waited for him. "You okay now?"

"I'm fine."

You were always fine in the infantry. You couldn't allow yourself to be any other way.

They'd been sent straight to the medics when they returned to BIAP, but before he and Brenner had gone, they'd stopped to look at the vehicle.

The inside was wrecked, but from the outside, there didn't appear to be any damage. A few pieces of shrapnel had struck Brenner's face and arms, and blood stained his pants. By some miracle, the wounds appeared minor.

Benjamin went back to that thing in the road as he stared at what had been his lifeline in this country. It had appeared to be an IED, and he'd tried to turn anyway. If he hadn't, would he or Brenner be dead right now? Would his platoon be sitting in their CHUs lamenting the loss of another man? Two?

"Are you all right, Benjamin?" Brenner put a hand on his shoulder.

Why anyone thought that gesture comforting was beyond him. "I'm okay, Sir."

Brenner pulled a camera out of his pocket. "Let's get a picture together. We can commemorate the day we got our purple hearts together."

Benjamin laughed at the silliness of it. The lieutenant spoke as if they were tourists, not soldiers. "I'm not sure I'll want to remember the day we got our enemy marksman badges, Sir."

Brenner leaned in close and took the picture anyway. He grinned at the camera in a way that made Benjamin want to roll his eyes.

"Great," Benjamin said as Brenner lowered the camera. "Now we'll always remember the day these assholes almost killed us."

Brenner examined the vehicle a moment longer, his smile fading. "It isn't the fault of these people that they're in this situation. Blaming all of Iraq is blaming the victim. The people who tried to hurt us are bad people, and we'll stop them, but never lose sight of the goal. We can make this country a better place for the people in it. They're human, just like you and me. Their lives are worth just as much as ours, and no accident of nationality can diminish that."

The silence of their dead vehicle deafened him as he turned over Brenner's words. It was nice to believe there was some deeper purpose here, but the truth was, no war was ever just. By its very nature, it couldn't be. Politicians would grab knives and duke it out themselves if it were. Instead, it caught people in the middle every time. Most of them poor. All of them hoping something better would come of it. It never did.

Pretending violence was the end to any problem was stupid. That all things could be broken down to their basest parts and destroyed. But Brenner hadn't meant it like that. Most other times his hearts and minds speeches would have been tiresome, but something about it then had been touching.

The medics at the aid-station looked them over. Brenner would need to go to a hospital to get shrapnel removed. Benjamin had a concussion. Word came down that they'd wanted to send Nelson home for his injuries, but he refused. He wouldn't leave his guys behind to fight a war while he sat at home wondering if they would make it. They'd started these fifteen months together, and they would finish it that way.

The headaches grew worse before they got better, but he was determined not to be a loser about it, even when Sutter, Ski, and Simmons noticed something was wrong with him for a few days after the incident. At one point, he'd stood up in the middle of the night, woken Sutter, and asked him if he was ready for breakfast.

"It's three in the fucking morning, dude."

When he woke up the next morning, he played it off as if it had been a joke. He wasn't laughing when his friends went on a mission the next day and left him behind. Without a truck, he became a useless appendage. Instead, the company sent him on details to keep him busy. Working with supply. Escorting TCNs. Anything so the manpower wasn't wasted.

But it was a waste. He'd been a soldier for a year now. Escorting people from Nepal around as they picked up trash was a chore for POGs, not for battle-experienced combat soldiers. It demeaned him, and he hated it.

And then one day, he was pulled off it.

"You gonna be riding with us today," Garcia said. "You ready to get back out there?"

The dread in his gut at the prospect of going out of the wire had never been there before. He'd always known it was dangerous, but the idea of doing it now brought to mind his brother's screams through the smoke and dust of his dead vehicle.

"Yeah, I'm ready, Sergeant. Been waiting for it. These details are for the fucking birds."

That night he went out for the first time in the weeks since he'd been blown up. Brenner was there too, and Nelson, though all in different vehicles. The back of the tanker's Stryker was as safe as his, but it didn't feel the same. That had been his. His Stryker. The home for his guys. He'd been the one to take care of it when it needed taking care of.

"Oh, shit, what is this?"

Platoon radio came alive as they stumbled upon the scene of what looked to be a murder.

"Green 7, Green 6. What do you want to do? Over."

Benjamin's morbid curiosity got the best of him. He tugged on Ski' pant leg as he stood in the hatch. "Let me check it out."

Ski stepped down and Benjamin replaced him, placing his nods over his face and casting the world in that familiar green glow.

Two dead bodies lay on the side of the road. Sometimes, the enemy used them as a place to plant bombs. They'd stick a few pounds of explosives under the bodies, and as soon as someone tried to move them, *boom!*

Benjamin stared at the corpses. They were older. Not like the two kids killed in Mosul what felt like a lifetime ago. Just two people. Victims of politics? Religion? The wrong place at the wrong time? It didn't affect him like it should, like it would a normal person. They'd turned that switch off in their brains. They had to if they wanted to survive a war.

Some said humanity had come far with civilization, but it hadn't. People found a mate, bred, acquired their comforts, and put their own needs before everyone else. That hadn't changed at all in ten thousand years. Morality had no place in nature. The meanest, hardest killers were the ones called the best of a generation. That was as true for the enemy as it was for them. The greatest sports stars were celebrated. The sharkiest lawyers. People weren't better than they had been when mankind lived in huts; they were just better at pretending. They no longer feared witches. Now it was terrorists. Not magic but bombs. Not demons but disease. If he met his caveman ancestors, he would see the same lights in their eyes he saw in the men around him. That little switch had been flipped for them too. The one that allowed them to see the blood and the corpses with nothing more than a shrug. There was no good or bad, just that path right up the selfish middle ground.

Benjamin ducked back into the vehicle. "All yours."

Ski took a moment to step back up. "You okay?"

"I'm fine. Just thinking how much it sucks to be them."

Ski smirked. "Not anymore." He climbed back into his hatch, leaving Benjamin alone with his thoughts.

The tanker vehicles didn't carry a full complement of troops like the infantry variant did. There was no room for it with the TOW[69] missile launcher in the vehicle. So while the others worked, he sat quietly in the back.

"Green 7, green 6. We'll call EOD and wait. Over."

He stared at the soft, glowing lights all around him. The FBCB2, bright as the sun in the dark space, lit the vehicle. The engine roared and whined at the front, and below him, the gears turned as they parked the vehicle to settle in and wait.

He was nervous, as he knew he would be on his first mission after being wounded, but there was a familiarity in it as well, as if they'd always been here. This had become their home. Not the United States, not BIAP. The back of these Strykers where they worked and lived. It was their sanctuary in a difficult world. It protected them from the Reaper, staying his hand time and time again on the battlefield.

Or maybe it hadn't. Maybe he and Brenner had died in that explosion, and only his ghost carried on. It felt like that, especially sitting in the back of a truck at night, listening to the million sounds of the technology around him.

---

[69] Tube-Launched, Optically-Tracked, Wire-Guided

The future, getting home, became a distant speck where the road met the horizon. Much closer was the soul-crushing certainty that he would die like his brothers. A lucky bullet to the brain. An explosion across the road. A bomb underfoot.

And so he sat there quietly, listening to the radio chatter of his closest friends, wondering what the future held for them, or if it held anything at all.

I am hurt, but am not slain;

I'll lay me down and bleed a while,

Then I'll rise and fight again.

— *Sir Andrew Barton*, English Folk Song

Afterword

This was a tough book to write. Not only because writing about my friends and myself was odd, but also because reliving a few sections was… well, not my ideal Sunday. Never before have I had to set a book aside so much while writing it.

I'd hardly consider myself a victim though, and I think you'd find that to be true of most of the people presented here. We're not broken, and even someone who has experienced the low points seen in this book probably wouldn't call themselves that. I don't. This book is just a small window into a life I hoped to shine a light on. An honest example of the complicated world that war creates for veterans and survivors. Not charming, but honest.

But hey, shit happens, right? Several of the people in this book and I have had issues crop up. Most of the experiences presented here are personal ones. Most everything is as true as it needs to be to get my point across.

It's hard to talk about. Both my upbringing and my military rearing taught me to hide hurts. Keep it bottled up until you can't take it anymore. But when your boiling point is high, you can be cooked before you reach it. I can count the number of times I've had a complete breakdown on one hand, but PTSD doesn't always work like that. As I said in the book, PTSD colors the way we see the world. It colors our actions. It changes us, inch by inch and mile by mile.

It took me a long time to realize it didn't make me weak to have these problems, and it took even longer to accept it. The idea that strong people should be bulletproof is flawed. It's tied up in this idea that a man's man is someone who can't feel pain, and that isn't realistic. Such a person does not exist. So I wanted to write about what those supermen look like under their skin. What they think about when they've hung up their gear and their medal rack is collecting dust in a box somewhere.

I think I did a fair job. It's tough to be critical of the military machine and the government they work for without being overly critical of the soldiers themselves. As I stated in the book repeatedly, I really do think soldiers are the best of a generation. That doesn't put them beyond reproach, of course. I think I made it clear that I feel as responsible for the current situation in Iraq as our elected officials should. I've always hated it when people say, "I support the troops, not the government." It ignores the fact that the troops aren't children, and that we signed up of our own free will. It's a nice way of saying you think them too incompetent to hold them responsible for their actions, and I don't appreciate it. I'm sure many other service members and ex-service members would agree.

But I'm getting off topic. I wanted you to have a glimpse into what it was like to be a soldier, during and after. I wanted you to see how insidious the road to mental instability can be, and the kind of thoughts that keep some of us up at night. I also wanted you to see that not everyone travels that road, and that being a victim of illness doesn't make one dangerous. Both of these points are important too.

As for everyone else in this book? They're all alive and well. I keep in contact with most of them. Two other people from Charlie

Company that I didn't mention in this book have gone down the river for murder. Another friend was killed while contracting overseas. A few have gone in and out of VA hospitals to "get their head on straight," as Allen would say. Most of them have gone on to lead normal lives as either soldiers or civilians.

For all you know, some of those Charlie Company killers could be your co-workers, your neighbors, or the man next to you on a bus. They are any one of a million people who went off to war, came back different, and then settled into a life that feels like it never quite fits right afterward.

We're everywhere. We're all around you. You don't see us most of the time, and I for one am okay with that. But try not to forget what we did, and that for a time we stood as soldiers. Remember that if you have strong feelings about the situation in the Middle East, many of us do as well. Remember the people whose countries were ruined, and remember that some of us shoulder the responsibility for it.

Remember, even when we take off the uniform, a part of it stays on.

Always.

-Al Barrera
May 2017